WELDING

A Management Primer

Employee Training Guide

By

Robert O'Con

INDUSTRIAL PRESS INC.
New York

Library of Congress Cataloging- in- Publication Data

O'Con, Robert L.,
    Welding : management primer and employee training guide / by Robert O'Con
        p. cm.
    ISBN 0-8311-3139-X
    1.Welding I. Title.

TS227.028 2000
671.5'2—dc21   00-039527

Industrial Press, Inc.
200 Madison Avenue
New York, NY 10016-4078

First Edition, May 2000

Sponsoring Editor: John Carleo
Book & Cover Design: Janet Romano

Printed in the United States of America

10 9 8 7 6 5 4 3 2

As always, for Lois

Other technical works by the author:

Welding Practices and Procedures (with R. Carr)
Prentice Hall, Englewood Cliffs, NJ, 1983

Metal Fabrication: A Practical Guide, 2nd Edition (with R. Carr)
FMA, International; 1999, Rockford, IL (1st Edition, Prentice-Hall, 1984)

Statistical Process Control:
A Training Manual and Implementation Guide
Fabricators and Manufacturers Association, International;
1997, Rockford, IL

SPC for Managers and Process Operators (video series)
Delmar Publishers; 1994, Albany, NY

Table Of Contents

Section 1 Welding Management

Section 2 Welder Training

Section 3

Acknowledgments

Critical to the writing of this text is the wealth of welding technical information and data produced and made available through The Lincoln Electric Company, The Hobart Institute of Welding Technology, and the American Welding Society. For the reader of this work who wishes to gain an in depth knowledge and awareness of the many facets of the welding industry, two texts can be highly recommended. They are Howard B. Cary's Modern Welding Technology (Prentice-Hall) and The Procedure Handbook of Arc Welding published by The Lincoln Electric Company of Cleveland, Ohio. Many of the data sheets used in this text are reproduced with their kind permission. For immediate reference and technical help, The Hobart Institute can be reached on-line (www.welding.org).

Photographs appearing in the text were taken by the author at the welding shop of the Herkimer County (NY) Trade and Technical Center. Thanks are extended to the instructor, Mr. Bob McGough, for his cooperation and to the center's director, Ms. Grace Dady.

Thanks also to my wife, Lois, for her energy and competence in typing the first drafts and the completed manuscript. Finally, thanks to the editorial and production staff of Industrial Press for their encouragement and professionalism throughout the publication process.

Preface

Why do our aluminum weldments develop cracks? How can we determine the true cost of our welding operations? Why do my welders insist upon the purchase of a brand of 6010 electrodes that cost a dollar a pound more than a reputable and well known brand with the same AWS designation? Why are my welders having difficulty in passing their D.O.T. Low-Hydrogen Re-Certification tests?

This book is written, in part, for the non-welder supervisor, manager, or shop owner who in the face of voluminous amounts of literature and data, needs to make informed and decisive decisions in order to implement and maintain a profitable welding operation or department. The text is also addressed to the apprentice/trainee and the working welder who more than ever before is asked to make autonomous and managerial decisions in every day work situations.

The first part of this work (Section I) is devoted to answering not only the questions asked above, but many more similar questions that will aid management in the supervision and control of their welding operations. Although much of the industrial and metal fabrication environment is highly automated, even robotically controlled, the individual welder and the manipulative welding processes he or she dominates represent an anomaly in an otherwise computer controlled manufacturing world. Yet the skill and often inventive near genius that an experienced and educated welder can bring to bear on the job can mean the difference between profit and loss, production quality work and endless rejects and loss of customer confidence. Specifically, this book is written to inform and educate non-welding personnel in the mysteries, methods, and trade secrets of the experienced welder. Additionally, chapters devoted to metallurgy, blueprint reading, and welding symbols, along with detailed process descriptions will enhance management's ability to make informed decisions on purchasing, supervision, and implementation of a variety of manual welding processes.

Section II is a complete curriculum for in plant instruction in the basic manipulative welding and cutting processes. The format allows the trainer to systematically present welding theory and practice to the student and to customize the instruction for any specific production objective. The aim here is to not overload the student (and the instructor) with amounts of data and theoretical material that, while ultimately of importance, does not directly and immediately lead to productive work and proper job performance.

The book concludes with a glossary, extensive data tables, and other references.

Robert O'Con

S E C T I O N

1

WELDING MANAGEMENT

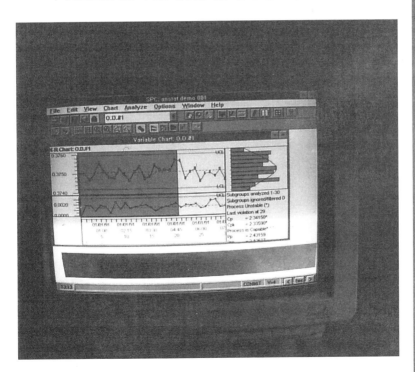

CHAPTER 1

WELDING FOUNDATIONS AND INDUSTRIAL APPLICATIONS

Welding is not simply sticking two pieces of metal together, but a total science and a primary method of construction and manufacturing used throughout the world.

Modern welding evolved from the ancient art of blacksmithing, which has been practiced for more than 2000 years. The two major processes, electric arc and oxy-acetylene, were developed almost simultaneously during the latter part of the nineteenth century. Two European scientists, Nikolas DeBernardos and Stanislav Olszewski, experimented with using a carbon electrode for fusion welding and were awarded the first electric welding patent in 1885. Development of the basic oxy-acetylene process was made in France by the chemist Le Chatelier in 1895.

Welding technology developed at an accelerated pace throughout the early 1900's. The two major advances that provided this rapid growth were: (1) the development of a safe method of compressing acetylene and oxygen into cylinders; and, (2) the introduction of the first coated arc welding electrode in 1912 which, for the first time, allowed welders to produce a high-quality electric arc weld. Prior to this time, electrodes were bare rods that produced weak and porous welds. The oxy-acetylene process was further advanced by the invention of efficient cutting torches, which made it the most practical method of severing iron and steel.

World War II was to be the final crucible from which welding would emerge as the foremost method of joining metals. The early 1940's saw the introduction of the gas tungsten arc welding (GTAW) process. It was developed initially for the aircraft industry because of the need to efficiently weld aluminum and magnesium. In 1948, the gas metal arc welding (GMAW) process was introduced and proven to be the fastest, most efficient method of joining thick sections of aluminum and other nonferrous metals. With refinements in power supply design and more reliable wire feed systems, the GMAW process became capable of producing consistently sound welds in all metals commonly fabricated in industry. When pulse technology was added it enhanced out of position welding capabilities and enabled the routine production welding of a variety of metal thicknesses. Programmable modules later became part of the power supply, permitting rapid set-up of established weld parameters. Oxy-acetylene cutting has been eclipsed by plasma and lasers. Use of the laser is generally limited to automated operations, and manual plasma applications, especially in cutting, are finding everyday applications in the shop.

Today it is the exceptional industrial process that is not facilitated, or in some way aided and abetted by welding. Meanwhile the construction industry has adopted welding as its primary method of joining, leaving behind almost complete dependence on riveting and bolting. Pipe joining is another beneficiary of welding technology. Joining pipe with assorted threaded connectors required wall thicknesses beyond that required by normal use, leaving interior surfaces that were irregular, reducing flow efficiency in the pipe. Welding not only eliminates such problems, it is highly productive in the on-site construction of a variety of piping systems.

Machine building and other industries that relied on large and intricate castings now use welding to reduce weight and facilitate more freedom and flexibility in design. The old foundry rules for uniform and minimum thickness are not needed for welded fabrications. Welding is also used to create special metal overlays that protect and conserve other materials from wear and tear and from corrosive materials and environments. The medical, food, and chemical industries also rely heavily on welding technology.

In providing these and so many other benefits, welding, ihas become very complex and technical. It requires not only specific knowledge but for the manipulative processes, considerable skill in order to gain the best the discipline has to offer. The list of advantages of welding include:

- ◆ the lowest cost joining method
- ◆ creates joints with high strength to weight ratios
- ◆ applicable to all commercial metals
- ◆ highly portable in most cases
- ◆ virtually unlimited design flexibility

On the down side, much of welding depends heavily on the human factor. There is also the need for inspection procedures that are often highly technical and complex . Welding and its applications are often restricted by specifications and codes that create an involved paper trail to show compliance. Still, these limitations can be overcome by good supervision, adequate quality control procedures, and by sufficient levels of education and training on the part of welding personnel.

CHAPTER 2

DETERMINING PROCESS APPLICABILITY

From the management perspective, the bottom line that would determine a welding process' applicability is cost per lineal dimension of weld. Yet arriving at that determination requires a body of knowledge and an awareness of both the physical plant requirements and the training and/or skills involved. This chapter will address those issues and help make possible an informed and considered judgment.

Before going on, there is an extension to that sought after bottom line. Weld quality and end use serviceability is an issue which is a two-edged sword. Swinging in one direction, the quest for quality can add additional costs that may not be merited. While in the other direction, poorly conceived shortcuts and economies can result in service failures, expensive rework, and possible legal issues. Specifically, it is the inappropriate and improper application of a joining process that will most often lead to excessive costs. From a broad perspective the three thermal processes that can complete a metal joint are soldering, brazing, and fusion. Each of these will be explored as to their potential value in a production situation. The actual instruction in the manipulative aspects will be covered in Section II.

Soldering

Soldering is a relatively low temperature adhesion process utilizing a filler material consisting primarily of tin, lead, and/or antimony. With variations in content, solder melting temperatures can range from 100° - 500°F and have tensile strengths in the neighborhood of 5,500-6,000 psi. Resistance to corrosion and electrical conductivity are other considerations in choosing a specific alloy. See the appendix for a compilation of common solder materials. Soldered joints are often supplemented by a mechanical joint connection such as slots and tabs, hemmed seams, or other geometric configuration. This is needed to insure the joint's integrity under vibration and stress.

Brazing

Brazing is a high temperature joining process approaching in some cases 1,800°, but still below the melting points of some of the materials commonly brazed. Generally brazing occurs at 1,300° to 1,600°. At those temperatures some surface alloying may take place between the brazing material and the base metal (braze welding). For the brazing of steels and cast iron, filler rods are made of alloys containing specific amounts of brass, copper, bronze, and lesser amounts of silicon, lead, and silver. In aluminum brazing, filler material will be alloyed with small amounts of silicon added to mostly pure aluminum stock. Successful aluminum brazing is limited to certain alloy groups, principally in the 6000 series

Silver brazing, sometimes inaccurately called silver soldering because of the way the braze material flows, uses filler material consisting of 45 to 65 percent silver. Silver brazing is used to effect dense, high quality joints in stainless, brass, and copper alloys. Such joints will also exhibit excellent electrical conductivity. The strength of silver brazed joints is significantly higher than similar tin/lead soldered joints. Both soldering and brazing require the use of specific fluxes to prepare the joining surfaces and improve the flow characteristics of the filler material. In general, soldering fluxes of the inorganic type are acid base, while milder fluxes are resin based. When brazing with the brass/bronze filler materials, a borax based flux is used. The silver brazing materials require a fluorine-based flux. There are specific flux formulations for a variety of applications including the joining of dissimilar metals. For general torch brazing of steels, prefluxed rods are popular.

Soldering and brazing are accomplished manually by hand-manipulated, gas-fueled torches or electric torches and irons. Higher production methods require the use of ovens or induction heating coils and elements. Prefluxed parts are assembled with preformed solder or braze filler materials in the form of rings or wafers placed on or between mating parts. Dip brazing and soldering is accomplished by "dipping" pre-assembled and pre-fluxed parts in baths of molten braze filler material. Dip brazing requires carefully controlled conditions and is usually performed by vendors whose facilities can include assembly and heat treating capabilities. All brazing and soldering processes require post-cleaning methods to remove residual flux materials.

Welding

Welding involves melting or fusion of two pieces of metal and usually, but not necessarily, the addition of a third filler material. The source of welding heat may be an oxy-fuel gas or an electric arc. The many types of fusion welding processes have been defined and categorized by the American Welding Society. Brief discriptions of the most commonly used, along with a brief description and their application, will follow. To keep within the scope of this text, the list is condensed to include those processes that are manipulative in nature and are more likely to be found in general use by both large and small shops and factories. The more exotic and "hi-tech" processes such as laser and electron beam welding are addressed in another chapter. Each process to be described will be preceded by its AWS designation.

OAS - Oxy-acetylene or "gas" welding is a manipulative rod and torch process which finds wide use in maintenance and repair work. As a production process its use has declined since W.W.II. Yet gas-welding is still a viable welding process for low carbon, mild steels less than 1/8 inch thick. Oxy-acetylene welding is well adapted to the joining of small diameter black iron pipe for gas lines. Other uses include cast iron welding, brazing, and flame cutting. Soldering is accomplished with acetylene, mixed in the torch, with ordinary air drawn from the surrounding atmosphere.

SMAW - Shielded metal arc welding, often referred to as "stick" welding, is used for all kinds of fabrication on a variety of metals with thickness usually of 1/8 inch or greater. SMAW is the most basic and economical of the electric welding processes, and uses a consumable electrode as a filler material. For the welding of mild steel (containing less than 0.25 percent carbon) the electrodes used differ only in their operating characteristics and recommended current type, which are determined by the chemistry and method of application of their flux coverings.

GTAW - Gas tungsten arc welding, once known by the proprietary name of "heli-arc," is now generally referred to as TIG. The process is used to fabricate many types of alloys, particularly stainless steel and aluminum, and is most efficient for welding material less than 0.25 thick. GTAW uses a non-consumable tungsten electrode whose heat producing arc takes place within a shield of inert gas of either argon or helium. Filler metal in the form of a bare rod may be added as in gas welding. The process can also be effectively automated. TIG filler rods are selected by alloy type and/or compatability with the base material and range in diameters from .030 to 0.125 inch and sometimes up to 3/16 or 1/4 inch.

GMAW - Gas metal arc welding is generally called MIG welding and is some-times called "fine wire" welding or by several other proprietary names. This method is a production alternative to SMAW and TIG welding. It is a semi-auto-matic welding method in which a coiled, consumable electrode is fed into the weld puddle within a shield of inert gas. MIG is used on many alloy types and is often automated or used with robot welding cells. Two distinct advantages of the MIG process are that the weld deposits are "low-hydrogen" in chemistry making the process more applicable to a wider variety of fabricated metals. Secondly, a wider range of material thicknesses can be welded with a single wire size. Typically, 0.030" or 0.035" diameter wire can effectively weld material from 1/16 to 1/4 inch thick. Depending on filler material alloy type, available wire diameters range from 0.020 to 0.125 inch thick.

FCAW - Flux cored arc welding is a variation of the basic MIG process. As its name implies, the consumable, coiled electrode contains a granular flux within its hollow core. The process is particularly suited for high deposition welding of steel plate thickness, and it also has some out-of-position capability and can be automated. Flux cored electrode varies not only by classification, but from man-ufacturer to manufacturer. Careful selection should be based on data and rec-ommendations supplied by the particular manufacturer.

PAW/PAC - Plasma arc welding is related to GTAW. Whereas the arc column produced in GTAW is conical in shape, the plasma arc is more cylindrical and thus more focused on the work surface. PAW is generally used to weld very thin materials, but plasma's more common application is as an alternative for oxy-acetylene flame cutting. PAC is particularly used on stainless steel and aluminum which cannot be cleanly cut using OAW. As a manual or machine guided process, PAC has virtually no metallurgical effect on the material and edges to be welded often require no post-cleaning or additional edge preparation.

Design for Welding

In the early years, as welding began to be seen as a viable alternative to other methods of joining (rivets, bolting, etc.), its acceptance as such was impeded by a failure to recognize the need for joint designs that were more suit-ed for this fusion of metals rather than the mere assembly of related parts. A very simple demonstration of this fact can be seen in Figure 2-1. The common sar-dine can shown is opened by pulling the tab, which causes stress along a line that has been embossed on the cover. Failure analysis studies have shown that whenever there is an abrupt change in grain direction, a weakness and the potential for fracture occurs at that point. Thus the sardine can opens neatly and smoothly along the embossed boundary, which in fact, is an interruption in the material's grain structure and direction.

FIG 2-1 The opening of the embossed cover along predictable lines is dependent upon a weakness caused by an abrupt change in grain direction.

Figure 2-2 sets up the same situation on two members overlapping each other in preparation for riveting or bolting. If a joint of this type and configuration is welded, a potential for failure exists at the points indicated. Of course other factors including the relative thickness of the members and the type of stress and loading the joint may be subjected to would all play a part. However, figure 2-2 also shows the simple welded alternative and how it results in the elimination of grain structure deformation. These examples do not even scratch the surface of weld design theory and practices, but their purpose is to emphasize that the successful application of welding is critically dependent upon sound mechanical design and intelligent joint geometries.

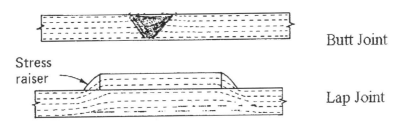

FIG. 2-2 Similar to the weakness in the embossed cover is the stress riser created when overlapping joints are welded. The simple alternative of the butt joint not only saves material, but could be stronger under the right load conditions.

The Cost of Welding

The actual cost per lineal dimension of weld can be determined by equating a number of factors in a variety of formulas. The principle factor is always time, time per weld, time per weld increment, or time to complete a part, with time being the basis for the ultimate calculation of the cost of labor. To this, caculation must be added the cost of filler materials, gases, fluxes, etc., joint preparation (if any), and finally, weld dressing and clean up. The cost of overhead must also to be factored in. Overhead would include costs of power, ancillary labor, and accessory tools such as chipping hammers, wire brushes, safety equipment and other needed tools. Whereas most shops or departments in a larger firm will have calculated their "shop time" as an overall hourly rate, welding shop time usually amounts to a higher than average cost than any other process activity. In particular and in addition to the cost of consumables such as gases and electrodes, power consumption is a significant cost factor and could increase the basic shop rate by 20 to 30 percent.

For welding and other joining processes, the following areas indicate where savings and significant cost reductions are possible.

Design:

- ♦ Incorporate formed or bent shapes where possible. Try to utilize the standard shapes and sizes of readily available mill products.
- ♦ Work with weldable and easily worked materials (refer to metallurgy in chapter 11) that do not require pre- or post-heat treatment operations.
- ♦ Reduce the cross sectional area of welds wherever possible. Better to have two smaller grooves on both sides of a plate than a single larger groove that will result in higher deposits of weld metal, excessive heat, and distortion (Fig. 2-3).
- ♦ Carefully analyze the size and amount of fillet welds and other welds. Doubling the size of a fillet quadruples the cross section and weight of the deposited metal. Interpret and follow weld symbols carefully. Welding symbols are addressed in Chapter 9.
- ♦ Pay attention to weld placement and joint accessibility.
- ♦ Make intelligent use of jigs and fixtures.

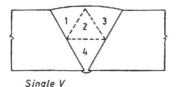
Single V

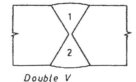
Double V

FIG. 2-3 A double groove not only reduces the amount of weld needed to complete the joint, but also helps to control distortion.

Procedures:

♦ Detailed job sheets, which convey established weld parameters and proven set-ups and sequences, are always useful

♦ Use maximum deposit rates afforded by the appropriate process selection, electrodes, and other filler material.

♦ Select a process with operator skill, fatigue, and training in mind.

♦ Pay attention to filler metal packaging. As in most other products, more is more economical.

♦ Be aware of electrode and rod wire stub loss. Normal stub loss in SMAW is 20 percent. Even at $1.60 per lb, that's 32 cents being swept off the floor.

♦ Join sub-assemblies first rather than "welding from the ground up."

♦ Use procedures to minimize distortion. See chapter 12.

♦ Utilize positioning equipment where possible, avoiding vertical and overhead welding particularly.

♦ Check equipment functioning and ground connections. "Hot spots" in power cables degrade equipment and waste power.

♦ Provide for operator comfort and safety. Good lighting and adequate fume extraction are essential. Encourage and facilitate good house-keeping. A clean shop is a safe shop as well as a more profitable one.

Related Manufacturing Processes:

♦ Strive for accurate part sizing to effect good joint fit-up and easy fixturing.

♦ Avoid designs that require excessive grinding or machining for edge preparation.

♦ Be diligent in pre-cleaning and degreasing operations.

♦ Dress welds and joints leaving radii and rounded corners wherever possible. Avoid the creation of sharp breaks and notches, which could be the beginning of a crack or fracture.

Computers in Welding

As the use of the PC has become more evident in the factory as well as the home, software designers have developed computer programs for virtually every aspect of manufacturing, including welding. Data bases have been developed to facilitate and aid in base metal selection, the selection of consumables by code requirement and manufacturer's recommendations, process and procedure selection based on joint geometry and service requirements, actual process operational control, inventory control of consumables, and for testing and quality control functions.

In this last respect, the storage and retrieval of quality related data is a critical issue in ISO certification and compliance. When such data is related to process improvement and process capability documentation, SPC software developed specifically for welding is especially useful. Chapter 7 presents an overview of SPC.

CHAPTER 3

THE NON-MANIPULATIVE WELDING PROCESSES

The welding processes described in this chapter are those that are operated by production operators trained specifically in their use. Although no particular manual skill is needed to make a weld, it is necessary for the operator to understand some of the theory involved. The choice of these processes is dictated most commonly by the level of repetitive quantities required and also by the nature and geometry of the parts to be joined. Compared to equipment used in the major manipulative process, these often called "high tech" joining processes are not only expensive in initial outlay and installation, but are often required to be subsidized by other plant operations.

Resistance Welding

Resistance welding is based on the heat generated by the resistance of the workpiece to a current passing through it. The combination of pressure and time duration, causes the area to be joined to be heated to a plastic state and forged together. The most common form of this method is known as "spot welding," however the same principle is the operative in flash welding, percussion welding, projection welding, upset welding, and resistance seam welding. Most commonly fabricated metals, including the precious metals, can be resistance welded.

Solid State Welding

Today these processes include some of the very oldest as well as some of the newest processes used in industry. They include cold welding, diffusion welding, explosion welding, forge welding, friction welding, hot pressure welding, and ultrasonic welding. Some of these processes offer the distinct advantage of producing no heat effects on the base metals because no melting or weld nugget forming is involved. As in resistance welding, time, temperature, and pressure are factors. However, the time is often measured in microseconds. Other instances call for hours of controlled temperature gradations. Solid state welding is particularly useful in joining dissimilar metals.

Electron Beam Welding

The electron beam process is a highly complex and expensive welding operation. Heat is generated by a concentrated beam of high velocity electrons impinging upon the work surfaces to be joined. The primary advantages of electron beam welding are its ability to produce deep welds with depth to width ratios of 20 to 1 or greater, and in welding the refractory and reactive metals that have a high affinity for oxygen and nitrogen, which will degrade the welds made on such materials.

Both low voltage (30,000-60,000) and high voltage (100,000 plus) methods are in use. Most electron beam welding is done in a vacuum although newer developments have reduced the need for extremely high vacuums and some allow for only atmospheric pressure during welding. Almost all metals can be welded with the process. Best suited are the super alloys, reactive metals like titanium, and the stainless metals. The mild steels are problematic and are not generally welded with this process.

Laser Welding (LBW)

The welding characteristics of the laser beam and the electron beam are similar. The laser is used to weld the more common metals as well as the specialty and more reactive alloys. Very high depth to width ratios are obtained with the laser. While the laser finds wider use in cutting, being often combined with CNC punching machines, laser welding continues to become more popular as initial costs decrease.

Thermit Welding (TW)

The TW process first appeared in 1900 and was once known as "cat welding" when applied to the joining of continuous rails. It is a gravity method where molten filler material is poured into a mold which is built around the parts to be joined. Large castings, rebars, and larger crankshaft components are typical applications. TW is a relatively inexpensive process considering the overall cost of the parts to be joined. A variation of TW is electroslag welding which allows the welding of heavy structural members in the vertical position. Although a distinct welding procedure, Thermit Welding resembles a casting process in some respects.

Robots and Other Automation

Like many other manufacturing processes, welding lends itself very well to automation. The GTAW process can be modified to feed coiled filler wire into the weld, while a tungsten electrode, mounted on a carriage, traverses a joint seam which is clamped to a mandrel or other back up device. The GMAW process is used extensively in robot controlled welding cells. Computer programmed, multiple, and repetitive welds are routinely accomplished.

Submerged Arc Welding (SAW)

The mostly automated SAW process utilizes a bare wire electrode fed into an arc that is covered by a blanket of granular flux. The flux covering obscures the arc light, eliminates spatter, excludes the atmosphere, and insulates the cooling weld.

The process is most often used for the repetitive welding of heavy structural members, railroad equipment, and larger-diameter heavy-wall pipe. Typically these applications require high amperage and the deep penetration that the process affords. Generally, the members to be joined are arranged to move or rotate under stationary welding heads.

During the building of the Alaskan pipeline, the 40-foot sections of 48-in. diameter pipe were joined with SAW into single 80-foot sections before they were set in place, either in trenches or on above ground supports. The "double jointed" sections were then manually welded with SMAW. The Alaskan pipeline extends from the North Slope on the Arctic Ocean, 800 miles south to the terminal facilities in Valdez (Fig. 3-1).

FIG. 3-1 The Alaskan pipeline runs over 800 miles from the north slope on the Arctic Sea to Valdez on Prince William Sound. This photo, by the author, was taken at Mile 562.

C H A P T E R **4**

SETTING UP THE WELDING DEPARTMENT

Welding Personnel

A welding department may be composed of just one or two welders or a dozen or more welders and permanently assigned welder helpers. Management must always be aware that in shop populations consisting of 20 or more persons and representing several trade areas, a distinct social structure and hierarchy will exist. Because of the arcane and specialized aspects of their work, welders are often isolated from the main stream social current in the shop, much like the inspection and quality control departments. Welders are often thought to be difficult and demanding to work with. If they are, it is because, when difficulties with the welding operation arise and rejection rates increase, the blame is nearly always directed at the individual welder. Certainly the blame for a poorly executed weld that leads to a reject should justly be placed on the welder or welders responsible. However, like a musician, a well trained welder will continue to improve and refine his or her skills for as long as they are engaged in the work. Whereas personal and personnel issues can affect performance, the vast majority of weld failure analysis reports find that it is anomalies in procedures, design, and weldment metallurgy that are the more common causes of weld defects and not always and simply, poor welder performance.

Weld Shop Layout and Support

The physical layout and set up of the welding department will vary to suit the process being used, but there are basic considerations that facilitate welder comfort for optimum performance, a smooth and efficient flow of work, and safety for welders and other personnel. In regards to the important issue of safety, specific process safety considerations are addressed in Section II.

Of primary importance are both lighting and fume exhaust. Good lighting ensures that assembly and fixturing prior to welding can be achieved accurately. With cables, hoses, and tooling being in constant use, adequate lighting will help prevent trips and falls. Besides being a prime concern with OSHA (Occupational Safety and Health Administration), proper fume and smoke removal is essential. Metal fume fever is a debilitating affliction caused by the burning of certain metallic coatings, principally steels that are galvanized or cadmium plated. Beryllium copper can give off toxic fumes that can prove lethal. Most welding processes generate large amounts of smoke that interfere with both breathing and vision. MIG welding (GMAW) when utilizing welding wire high in magnesium is particularly irritating to the eyes.

Where welding is performed in a separate walled-in room or area, efficient smoke and fume exhaust requires additional air intake equipment to facilitate the movement of air and increase the efficiency of individual welding station exhaust ducts. Welding stations in a large open shop area can usually get by with an exhaust duct in the immediate weld area.

Eye Protection

Individual welders working in close proximity to each other should be separated by opaque canvas curtains that have been treated with a fire retardant. To protect casual observers and other nearby workers, transparent vinyl screens are often used. These screens will significantly, but not entirely, reduce the arc glare of electric welding. The bright light of an arc can cause some real damage to the unprotected eye. However, getting "flashed" as it is called, referring to intermittent and occasional exposure, usually results in a "sunburn" on the eyeball and can be very painful. Treatment can range from applying clean cold water to the use of ophthalmic medications specifically used to aid in soothing and healing. Welder-helpers and assembly workers in the immediate area are usually the victims

Determining Welding Machine Size and Output

The largest single purchase in setting up the weld shop will be the welding power sources. Welding machines are sized according to their maximum amperage output, 100 amp, 200 amp, 300 amp, and so on. Critical to their cost is their rated duty cycle. Duty cycle is a product standard which defines how many minutes out of ten a welder can be operated at its full rated output before it begins to overheat and deteriorate. Well known brands are conservatively rated, and lesser known economy models may prove to be a poor investment.

In setting up the weld shop, thought must be given to the expected welding amperages that will be used in relation to the power supply's duty cycle. For example, if the production amperages are expected to be in the 140 to 160 amp range, a light duty 225-amp welder with only a 20 percent duty cycle will not be adequate for continuous use at those amperages. In contrast, a 300 amp, industrially-rated power supply with a 60-percent duty cycle at 300 amps would be virtually indestructible when used at 140 to 160 amps. Production welding requirements often include down time for set up, fixturing, and tacking. However, the large initial cost of the power source demands a careful review of the shop's work load, intensity, and welding parameters.

Electric Power and Compressed Gases

The welding shop power and compressed gas requirements also vary with specific process needs. Except where electric motor generator welders are used, welding input power is usually 220 volts, single phase. Multi-station shops with more and larger power supplies should be wired for 440-or 600-volt circuits. These higher voltages will reduce overall amperage consumption and voltage drop throughout the shop. When considering gas supplies, either inert for shielding or oxy-fuel (usually acetylene) for cutting or welding, multi-stations are best served by remote manifold systems. Piping of the needed gas to each station, eliminates the clutter and mess of cylinders and hoses on the shop floor. Additionally, savings in down time to change and handle cylinders and in reduced cylinder demurrage charges can be realized. Manifold systems also are an important safely step that can sometimes lower insurance costs. By locating gas supplies outside and a distance away from the actual welding operations, the potential danger to shop personnel is greatly reduced. Cylinders that must be kept in the shop and at or near the welding station should be secured to a wall or stanchion or mounted in a cylinder cart.

Work Flow and Weld Tooling

Weld shops or departments usually evolve as an afterthought or as a reactive knee-jerk to production realities. While most other departments and functions are planned with accurate foresight and much thought, the welding operations are usually left to develop by themselves with the welders making or requesting changes and alterations in the midst of actual production. Variations in weld shop design will always reflect the work being done, yet there are basic

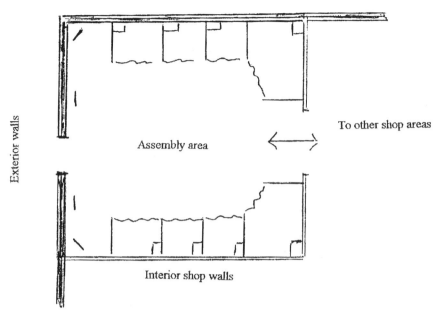

Exterior walls

Assembly area

To other shop areas

Interior shop walls

FIG. 4-1-a "Wall" arrangement for a weld shop. A number of individually divided and curtained welding booths are separated from the rest of the shop.

Surrounding Shop

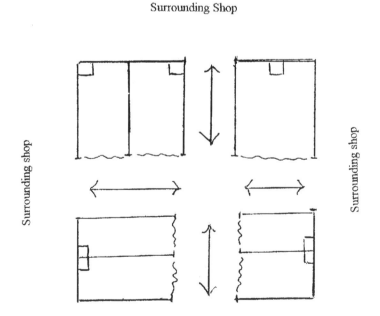

Surrounding shop

Surrounding shop

FIG. 4-1-b In the "island" arrangement, the welding area is partitioned off from the rest of the shop with a rigid and semi-permanent curtain wall. Here, several various sized individual welding booths contained within the island.

considerations that can be anticipated with an eye toward ergonomics and production efficiency.

The two basic weld shop arrangements are the wall set up and the island set up (Fig. 4-1a,b). As seen in Figure 4-1a, the wall arrangement facilitates distribution of electric power and gas and can also simplify exhaust systems. The center space can be used as an assembly area or a staging area for presenting work to the individual welders. Delivery of parts in and out of the area is easily accomplished on rolling carts, tables or fork lift pallets. Large weld platens or jigs can also be placed in the center. Serviced by longer welding cables and surrounded by portable screens, a shop within a shop can be created.

In contrast, the island arrangement, as shown in Figure 4-1b, makes welding the center of all ancillary operations including fixturing, pre- and post-cleaning, and any required grinding and weld dressing operations. Closed in by its surrounding wall of curtains, the island acts as a floor divider for the other departments in the shop and does not restrict access in and out of the weld area proper. In the island arrangement, electric and gas delivery systems are not as easily installed and may require the use of ceiling mounted drop lines and piping. However, the island set up can be easily expanded and is conducive to good house cleaning and maintenance.

Critical to efficient weld shop operations is an adequate supply of clamps. Other items include chipping hammers and wire brushes for stick welding (SMAW), diagonal cutting pliers and anti-spatter compounds for MIG operations (GMAW), stainless steel brushes, and a supply of stainless steel wool for the manual removal of surface oxides from aluminum and magnesium work pieces and bare welding filler rod.

TIG welding (GTAW) employs a high frequency starting current which facilitates non-touch arc initiation and stabilizes the AC current output. This high frequency current will sometimes arc between a poorly-grounded workpiece and a work table. To minimize aluminum workpiece damage caused by this arcing, TIG work table tops are made of aluminum plate which should be periodically wire brushed to remove oxide build up. This high frequency current may also interfere with closely located radio frequency devices.

Part Cleaning

An area devoted to part cleaning should be maintained near the welding operation. TIG and MIG welding are particularly unforgiving in respect to part cleanliness. Steel will most often require degreasing and aluminum will require a mild etching process to remove surface oxides. Rinsing and hot air drying can also be part of the cleaning sequence. Aluminum and other non-ferrous parts to be welded can be stored in plastic bags to retard the reformation of oxides.

Welding Consumables

In addition to gases, electrodes, and other filler wires, the weld shop inventory should include replacement stock for the following:

- Tungsten electrodes of various diameters (GTAW)
- Ceramic gas cups (GTAW)
- Tungsten collets and collet bodies (GTAW)
- Wire liners (GMAW)
- Wire contact tips (GMAW)
- Welding helmet clear lens replacements
- Gauntlet gloves (SMAW and GMAW)
- Light weight TIG welding gloves
- Leather welding garments, including jackets, bibs, sleeves, and skull caps

CHAPTER 5

DESIGNING AN IN-PLANT WELDER TRAINING CURRICULUM

The acquisition and training of new employees is one of the critical issues facing businesses as the twenty-first century begins. It is the rare employer who does not say that their business volume could at least double if they had the people. As older workers retire from the metalworking trades, industry looks to the schools, both public and private. While many private vocational schools can and have been supplying specifically trained workers, including welders, the vast majority of public high school students today show little interest in the metal trades. So it would seem that the most reliable source of welder trainees would be those already on the payroll who are working at entry level and semi-skilled jobs in various other departments. Women also represent a potential for welder trainees. Many welding processes associated with the aerospace and electronics industries can be effectively operated and performed by mechanically inclined women.

Except for the most repetitively intense production work, welders may be required to read and interpret blueprints, perform mechanical assembly, and use a variety of measuring tools such as verniers, protractors, squares and the like. Additionally, a working competency in basic computational math, including plane geometry, is valuable as a core discipline for trainees and apprentices to any metal trade.

On-the-job training (OJT) has always been the most effective in fitting new hires into the production scheme of things. Older line supervisors and foremen will recall the times when very green but fairly well educated and literate new employees were brought up to speed in a relatively short time. Unfortunately many of today's high school graduates lack the basic academic knowledge that facilitates effective on-the-job training. It is difficult to explain the nuances of DC and AC welding current to someone who has not grasped the concept of magnetism and polarity. Welding to print dimensions is a challenge to one who has not become proficient in simple computational math and the conversion of conventional point-to-point dimensions to datum and baselines. Given the need to supplement OJT with formal classroom instruction, it is necessary to select a method of delivery and decide who is going to provide the instruction. The following list outlines the various delivery systems available to the employer.

Training Methods

The central question to ask of any training methodology is whether it is effective and efficient. While effectiveness can directly improve performance and productivity, efficiency deals with the cost and the timeliness of the training. The two elements are at odds, requiring a trade-off that includes careful considerations in selecting any training method. Several learning tools can be considered when implementing training.

- ♦ **Computer-Based Training.** Computer-based training can be self-paced and accomplished with existing hardware. However, programs can be expensive, and the less interactive ones may involve simply reading electronic pages.
- ♦ **Video Training**. Video training can be effective when presented in small, specific segments. The ones that electronically reproduce a classroom setting can be less effective. New developments in interactive video, multimedia learning, and virtual reality training can be explored. However, the value of each has yet to be quantified, and they can be expensive.
- ♦ **External Colleges and Vocational-Tech Schools.** With the right industry input, these areas can be very effective, but with a high loss of employee productivity. Teletraining is somewhat more efficient but can present scheduling problems.
- ♦ **On-Site Consultant Trainers.** When specifically oriented to training needs, consultant programs can be an effective and efficient means of delivery. Cost can be a factor if the program is drawn out.
- ♦ **Prepared Courses.** The problem with prepared courses is that it can be difficult to motivate employees to use them with the right attitude and enthusiasm. Neither the employer nor the employee wants to build or work in a schoolhouse.

♦ **On-the-Job Training.** OJT is the most efficient in terms of cost and productivity. However, effectiveness suffers if the training becomes too informal or unstructured. On-the-job training is most successful when basic skills and knowledge are firmly established. When supplemented with short and intensive classroom instruction, this training method can be a model of effectiveness and efficiency.

If an in-plant classroom program is selected with classes held after normal work hours, line foremen, lead persons, and supervisors are excellent candidates for the role of instructor. In addition to getting to know their people, the instructor can tailor the instruction to the working realities of the department. A slight digression is called for at this point. To whatever degree the principles of total quality management are implemented within an organization, the role of supervision has and will continue to evolve over the coming years. Instead of being the pusher, the old authoritative foreman barking orders and controlling every aspect of the department is fading away. Industry is beginning to recognize supervision in the role of leader and job coach directing the activities of those in their charge. As employees become more self-directed and more directly responsible for their own job performance, supervisors will have more freedom to concentrate on process improvement as well as job enhancement and training. For the new welding trainee, formal instruction in blueprint reading, and shop math, including the basics of algebra, trigonometry, and geometry, is considered to be a sound investment, adding value to a firm's employees. Instruction in these subjects are to be found in Section II.

To instruct welders in the manipulative skills involved in welding, many training hours can be involved. Unless there are experienced welders on staff who can take on the responsibility of supervising the practice sessions, it will be more productive to enroll the trainee in a public or private welding school. Many public vocational schools or community colleges offer industrial welding classes. For more specific and intensive training, a private trade school would be able to get an employee up to "productive speed" quickly. In some instances, it might be feasible to bring in a private instructor on a consultant basis.

WELD TESTING AND
WELDER CERTIFICATION

This chapter will present an overview of a very large and comprehensive area of the welding industry. The purpose here is to acquaint the shop owner and manager with those aspects that will affect costs, quality, and the potential for legal liability.

The demand for high quality products in the market place, increasingly complex technology, and ever increasing competition for market share, make the issue of weld quality of prime importance. In certain fields, such as nuclear energy and aerospace, quality means 100 percent reliability of every single weld. In other areas, product liability is a major factor and is related to expected service use. Many industries have adopted rigid codes and specifications. These would include, besides the nuclear and military requirements mentioned, piping. ship building, bridge and building codes, along with the manufacture of storage tanks and other vessels. Yet, for many manufactured products, no codes or standards exist and as society and consumers become more prone to pursue their legal rights, manufacturers must seek ways to protect themselves.

The responsibility for producing quality welds can rest with many people from shop owners and supervisors to quality control and inspection personnel, and with the actual welders and welding operators. Two areas to be addressed here are weld procedures and actual welder performance.

As a vendor, a firm may be audited by the welding engineering department of a prospective customer. The vendor's ability to perform to contract specifications is determined by that audit. Weldment preparation, cleaning, fixturing, pre- and post-heat treatment facilities are typical concerns and of course, welder skill and ability as determined by some sort of test. For such a test, welder qualification and welder certification must be defined. Qualification is a measure of the ability of the individual welder to perform to an established standard, and is proven by taking and passing a prescribed practical welding test. Certification however is a written statement attesting to the fact that the individual welder is capable of welding to the standard in question. Certification implies that a testing organization, manufacturer, contractor, owner, or product user has witnessed the preparation of the test plates and sometimes the actual welding, has conducted the prescribed testing methods, and has recorded the successful results in the appropriate manner for the specific agency or jurisdiction. Welder qualification and subsequent certification may then be considered another cost of doing business. In most cases smaller vendors or individual job shops would be well advised to use the services of an independent testing laboratory. Although the initial cost may seem high, the establishment of an irrefutable paper trail can be very much worth the cost. Besides having all the right testing equipment and a knowledge of the various specifications, the testing organization can also supply the vendor with prepared test plates of known and guaranteed analysis.

Weld testing occurs in two basic forms, destructive and non-destructive. Most common certification tests are the destructive kind. Samples, called coupons, are cut from welded test plates to a specific size and subjected to guided bend or tensile strength test machines. In other cases the whole test plate or actual production weld sample may be subjected to various non-destructive testing methods such as X-ray testing (Fig. 6-1). Even after an initial

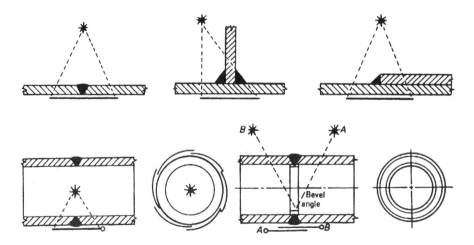

FIG. 6-1 X-rays are a common form of non-destructive weld testing. Shown here are some typical test joints and the arrangement of film and the X- ray source for each. (Courtesy Lincoln Electric Co.)

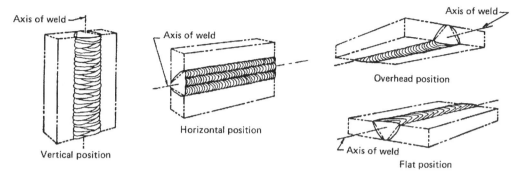

FIG. 6-2 Test welding positions. (Courtesy Lincoln Electric Co.)

certification test is passed, periodic check tests, usually every six months, may be required. Thus a continuous record of welder proficiency is established over the course of employment with a particular firm. Some forms of welder certification may be considered permanent in the sense that the records will follow a welder from one employment to another. However, many welders must be recertified upon employment with a new firm. Certifications are obtained in specific processes, specific alloys, and in some cases, in certain positions. Figure 6-2 illustrates the various designated welding positions.

For non-code welding and to verify the skill and ability of a newly hired welder, there are several easily administered tests that can be given. First and foremost and regardless of the process or alloy involved, is weld appearance. Figure 6-3 shows several properly executed welds. Each are evenly rippled, smooth appearing and gives evidence of satisfactory training and/or experience. After a few warm up beads, the applicant should be able to set the machine

FIG. 6-3 Weld bead appearance is a strong indicator of proper technique and equipment use.

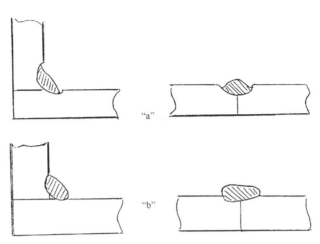

"a"

"b"

FIG. 6-4 Welds labeled "a" are "undercut" reducing metal cross section. Welds labeled "b" are "rolled over" resulting from poor technique and/or improper amperage.

controls for the sample weld and demonstrate the proper technique to achieve the required weld quality. Visual defects are shown in Figure 6-4. The weld labeled "a" shows excessive "undercut." The weld labeled "b" is "rolled over."

Beyond visual inspection several physical tests can be performed. For electric arc welding (SMAW), a simple butt and tee joint is welded on one side only (Fig. 6-5). If the welds pass a visual inspection they are next broken in half by placing the root of the joint in tension (Fig. 6-6). This is not a strength test but rather a simple way to inspect below the surface of the weld. The weld bead should crack, after some initial resistance, down the middle of its cross section

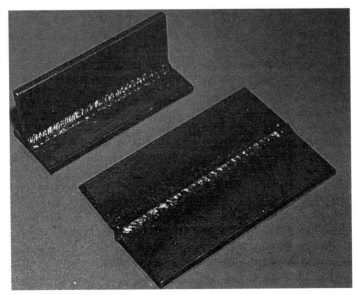

FIG. 6-5 Simple butt and tee joints can be used to test a welders basic skills.

rather than "pulling" out of one side or the other. The revealed weld metal (Fig. 6-7) should show evidence of solid weld metal casting, free of slag and gas pockets, throughout its length, except for possibly the beginning and end of the joint.

TIG and oxy-fuel gas welds, which are usually done on thinner materials, can be tested by cutting the weld samples into one inch portions and, as shown in figure 6-8, bent over until flat. Fracture, if it occurs, should be in the adjacent material and not in the weld itself.

Heavier weld samples can be sectioned and etched with an acid to reveal the weld density and any foreign inclusions. Figure 6-9 shows an alu-

FIG. 6-6 Test welds are broken by placing the root or opposite side in tension, as shown.

FIG. 6-7 The revealed weld metal should be free of gas pockets, slag, and show even penetration.

FIG. 6-8 A section of a thin sheetmetal joint folded over, again placing the root in tension. If a fracture occurs, it should be in the material and not the weld bead.

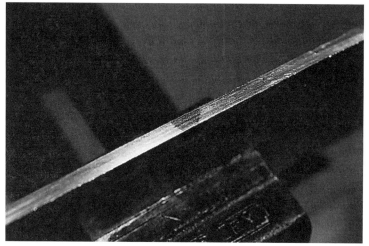

FIG. 6-9 Heavier test joints can be etched to show weld density.

minum weld coupon etched with a hydrofluoric acid solution. Steel coupons are etched with a ferric chloride solution. Other test methods that the individual firm can perform are the tensile strength test for thinner materials and the guided bend test for plate thicknesses. For test purposes, two pieces, each measuring 4 by 6 inches long, are butt welded together. Sheetmetal thicknesses are generally welded from one side with full penetration and reinforcement to the opposite side. The guided bend test is performed on 3/8-inch. thick plates of the same 4 by 6 -inch size. They are also welded with full penetration and protruding reinforcement on the opposite side. In each case the weld coupons must be properly prepared and the test carefully conducted to provide a fair and objective, as well as an accurate result. Coupons to be subjected to tensile testing should be cut to a width of one inch, and ground flush on both sides with the cut edges given a smooth radius. The tensile testing machine has opposing jaws that grip the coupon and are pulled apart hydraulically until a fracture occurs. A gauge indicates the tensile force applied at the moment of failure.

The guided bend tester (Fig. 6-10) bends the coupon over a specified radius, usually 3/4 in. On a coupon, 1-1/2 in. in width, approximately 60,000 lbs of tensile force can be applied to both the face and the root of the weld. This a force equal to the strength of the unwelded metal.

The American Welding Society publishes detailed information on a wide variety of testing procedures for the many agencies that require and use specific codes.

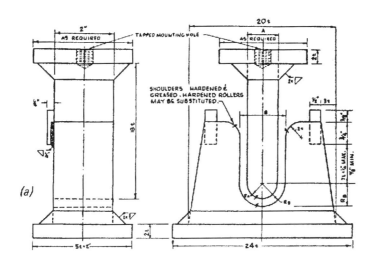

NOTE: "t" refers to specimen thickness
"t" for AWS test is 3/8"
"t" for API Std. 1104 is tabulated wall thickness of pipe

Jig Dimensions	AWS TEST For Mild Steel Min. Yield Strength-psi			API Std. 1104 For All Pipe Grades
	50,000 & under	55-90,000	90,000 & over	
Radius of plunger R$_A$	3/4	1	1-1/4	1-3/4
Radius of die R$_B$	1-3/16	1-7/16	1-11/16	2-5/16

FIG. 6-10 The guided bend test is a basic procedure for determining advanced skills for both steel plate and pipe welds. (Courtesy Lincoln Electric Co.)

CHAPTER 7

STATISTICAL PROCESS CONTROL (SPC) IN WELDING

SPC is finding ever-expanding use in measuring and predicting a wide variety of industrial process results. The application of statistical methods to welding processes is often ignored because of the amount of codes, procedures, and agency specifications that already govern most production welding operations. Yet that does not negate the fact that SPC can be very useful, just as it is in other industrial processes, in reducing variation in weld quality and in enabling continuous process improvement. SPC is not an inspection procedure, but rather a way of quantifying weld process capability and performance.

As companies, large and small, begin to incorporate the concepts of total quality management (TQM), quality becomes what customers say it is and not what suppliers and their inspection departments determine it to be. Put another way, while a supplier may boast of "a less than 1 percent rejection rate," increasingly sophisticated customers are learning that such boasts reflect the zeal and energy of inspection and rework departments and the scrapping of rejected products, rather than the actual productive capability of the supplier. Moreover, and in the face of global competition, customers are not willing to subsidize such zeal and energy through higher prices.

Across the industrial spectrum, the cost of ensuring quality through quality assurance procedures is said to amount to an average of 20 to 40 percent of final product cost. In welding, the figure may approach 80 percent. Thus it is obvious that competitive pricing can most effectively be realized by reducing the cost of quality. This reduction can be accomplished by continuous process improvement through the application of statistical methods. SPC provides for objective statistical expressions of process capability and process performance, generating the statistical references by which an increasing number of firms are evaluating their suppliers.

As part of this introduction to SPC and how its principles can be applied to welding, the critical difference between capability and performance must be defined. Capability is the highest average performance a process is capable of once all "special" causes of variation are removed, and leaving only "random" causes to affect process variation. In welding, such special causes of variation can include inconsistencies in pre- and post-heat treatments, joint fit-up and pre-cleaning abnormalities, variations in metallurgy, and even levels of welder skill and training. Once these special causes of process variation are eliminated or stabilized in some way, what remains are only the natural or statistically random causes. This leads to the definition of performance as being the day-to-day process output without the effects of the special causes of variation.

The statistical data that can be generated by any process, including a welding process, can be either a variable or an attribute. A variable is an observation that can be measured and will have a range of acceptable plus or minus values. Attributes, however, are counted rather than measured. There is no range of acceptance, only "yes-no" and "go-no-go" types of observations. As such,

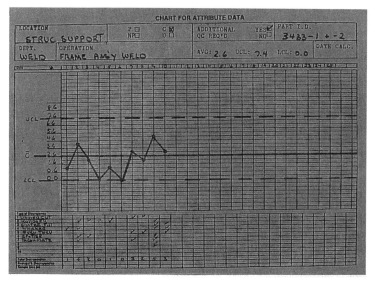

FIG. 7-1 An SPC attribute chart which statistically quantifies weld process performance over a period of time.

attributes are most often applied to welding processes. Both variable and attribute data is recorded on specifically formatted charts. The charts can be paper hard copies (Fig. 7-1) or computer generated with software programs designed expressly for SPC applications.

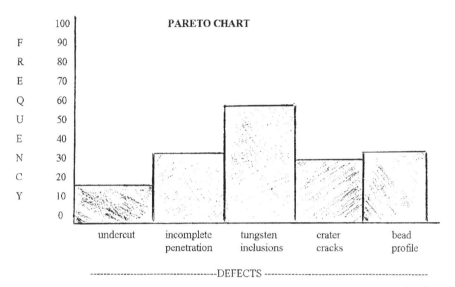

FIG. 7-2 The Pareto Chart graphically depicts recurring causes for part rejection. In this case it is tungsten inclusions found by X-ray inspection.

The Pareto Chart is one of the more simple charts to create and use and is very illustrative of the value of SPC in welding. The Pareto Principle states that if there are ten possible causes to production or quality problems, only one or two will be the most recurring and probable causes. The chart shown in figure 7-2 illustrates that principle. The horizontal line categorizes the different possible causes for a weldment's rejection. The vertical line records the number of each kind of defect found. In this sample, for a TIG welded part, the recurring causes for rejection are tungsten specks appearing on an X-ray of the fused area. While other causes do exist, it is now statistically dictated that management's effort to solve the problem be directed at the specific problem indicated and, for the time being, no time and effort be wasted on the other less probable causes.

There can be several reasons for "tungsten spitting" as it is termed. A careful examination of the contributing factors is the essence of continuous improvement. Included are excessive amperage, poor torch cooling, wrong type of tungsten, power supply wave balance problems, and welder dexterity and steadiness. A cause and effect or "fishbone" chart as seen in figure 7-3 could even reveal that the welder is drinking too much coffee during the day. The point is that SPC is more of a journey than a destination. And as a journey of discovery, the signposts of continuous process improvement are constantly encountered. It only remains for the welding manager or shop owner to do something about them. If all you do is record data without corrective action, then you are

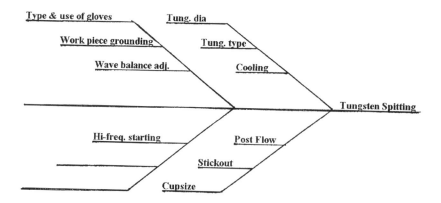

FIG. 7-3 The Cause and Effect or Fishbone Chart is used to pin down all the possible contributing factors which could cause tungsten spitting.

only recording history and historians are not well known for making money or even surviving.

This chapter was intended solely to provide a brief introduction to the use of SPC in the welding operation. The welding manager or shop owner would do well to study one or two of the many excellent texts devoted to the subject of SPC.

S E C T I O N **2**

WELDER TRAINING

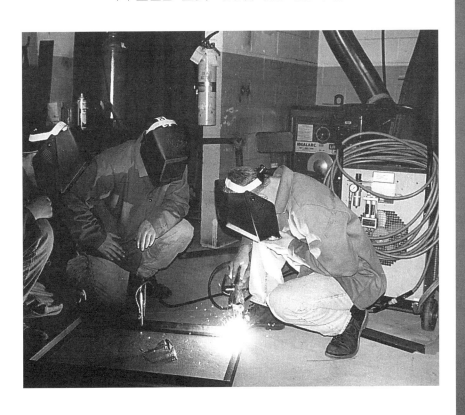

WELDING SAFETY

Welders work with potentially dangerous equipment. Although responsible companies in the industry as well as OSHA have made significant advances in welder safety, the major portion of responsibility still lies with the individual welder.

The purpose of this text is not only to help the employee to become proficient in the trade, but also to be safety conscious. Learning and practicing the following safety rules will ensure a safe, healthy career in the welding industry. Management's support of safety and provision of the tangible and philosophical foundations for a safe shop are well within the realm of rational self interest.

General Rules

- ◆ Remove all matches or other combustibles from your pockets prior to welding or flame cutting.
- ◆ Remove any rings from your fingers prior to arc welding.
- ◆ Do not pick up any metal in the welding shop without testing to see if it is hot.
- ◆ All injuries, no matter how minor, must be reported immediately.
- ◆ Clear your work area of flammable materials and unauthorized personnel prior to welding or flame cutting.

♦ Never attempt to lift heavy equipment or materials without adequate help.

♦ Flash goggles should be worn when assisting in a welding or cutting operation.

♦ Safety goggles should be worn at all times when working in an eye hazard area. Blindness is a terrible risk for a moment of carelessness.

♦ Protect your feet when welding or cutting. Wear leather boots that are at least ankle high.

♦ Always wear appropriate work clothing (long-sleeved cotton or canvas shirts, denims, coveralls, etc.) when performing welding or cutting operations.

♦ Never use equipment without first being instructed in its safe handling and use.

♦ Never weld or cut directly on concrete.

♦ Be aware of the hazard of toxic fumes. When welding or cutting metal that is galvanized or painted, clean as much of the heat affected area as possible and always work with proper ventilation.

♦ Never flame-cut or weld in or near an explosive atmosphere, which includes an oxygen enriched atmosphere.

♦ Never weld or cut on a vessel that has contained a flammable substance unless the vessel has been properly cleaned and rendered safe.

♦ Develop the habit of checking your equipment frequently. Make sure it is always sound and serviceable.

Arc Welding Safety

♦ Never attempt to perform or observe arc welding without wearing the proper welding helmet fitted with the proper shade (#10 minimum).

♦ Always be sure that your welding power supply is properly grounded to avoid accidental shock.

♦ Never attach the welding ground clamp to any pipe or vessel that contains flammable liquids or gases.

♦ Inspect welding cables frequently for cracks in the insulated cover. Repair or replace defective cables immediately.

♦ Wear protective clothing. Rays from the arc can cause severe burns on unprotected skin.

♦ Protect your eyes at all times. Prolonged exposure to the ultraviolet and infrared rays of the arc can cause permanent eye damage.

♦ When chipping slag, remember that it is hot enough to burn your skin and eyes. Always wear safety glasses and try to chip in such a way as to direct the slag away from you.

♦ The input electricity for your welding power supply can kill you. Never perform maintenance on your machine without first disconnecting it.

♦ Never touch an uninsulated part of the electrode holder, ground clamp, or welding cables.

♦ Never attempt to operate the polarity switch while the power supply is under welding load.

♦ Always warn those around you before striking an arc.

♦ Never weld in a damp or wet area without proper insulation against electric shock.

♦ Check your welding helmet for cracks and chipped or broken lenses. Replace defective lenses immediately and keep them clean.

Oxy-Acetylene Safety

♦ Never leave a lighted torch unattended.

♦ Never oil oxy-acetylene equipment.

♦ Treat gas cylinders with extreme care. They have the potential explosive force of a bomb.

♦ Always keep cylinders chained in the upright position.

♦ Always use a striker to light a torch, never a match or a cigarette lighter.

♦ Always wear approved goggles when using a torch (#4 lens minimum).

♦ Always leave the T-handle in place on the acetylene cylinder, and never open the cylinder more than one full turn.

♦ Be aware of what is around you before lighting a torch.

♦ Never stand in front of a regulator when opening a cylinder. Stand off to one side.

♦ Never use obviously worn and defective equipment.

♦ Make sure that the regulator adjusting screw is fully backed out before opening the cylinder.

♦ Always check the oxy-acetylene outfit for leaks prior to lighting the torch.

♦ Check hoses frequently for cracks and cuts.

♦ When making repairs, use only approved fittings.

♦ Never transport cylinders without safety caps.

♦ Always open the oxygen cylinder valve until seated in the full open position.

♦ Always bleed the system down and then release regulator diaphram pressure by backing out the thumbscrews counter-clockwise.

Protective Equipment

The welding helmet is constructed of fiberglass or heat-resistant plastic and will protect the welder's eyes from the harmful rays of the arc, and the head, face, and neck from arc burn and sparks. See Figure 8-1 for proper placement of the filter and cover lenses. Additional filter lens information is given in Table 8-1.

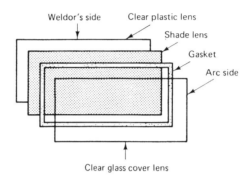

FIG. 8-1 The lens assembly shown will reduce double images and protect the expensive filter shade lens. (Courtesy American Welding Society)

Welder's gloves are made of leather. Although several styles are available, the gauntlet-type glove offers the most protection for hands, wrists, and forearms. Leather welding jackets and aprons are specifically designed to protect the welder from sparks and molten metal, particularly when welding or flame-cutting in the vertical and overhead positions. For TIG welding, special light weight gloves and slip-on protective sleeves to protect the forearms are available.

Additional Eye Safety Devices

Spectacle-type industrial safety glasses with protective side shields offer the best eye protection and should be worn at all times. Flash glasses are safety glasses with tinted lenses and side shields. They give the welder protection from arc glare and should be worn at all times in areas where arc welding is being done. A lens shade No. 2.5 is recommended. Welding goggles are designed to protect the welder's eyes from the glare of the oxy-acetylene torch and from sparks and molten metal. Most goggles use a 50-mm round lens. A lens shade No. 5 is recommended for most oxy-acetylene welding and cutting operations. Safety face shields are constructed of thick, shatter-resistant, clear plastics. The shield offers maximum protection to the face and eyes and should be worn when operating power grinders.

The responsibility for the implementation of safe work habits both in welding and the operation of related machinery ultimately lies with the individual worker. Modern welding equipment has many built-in safety features. Unless the welder learns how to utilize these features, they serve no purpose. Items of particular significance are the high input voltages of the power supplies, use of compressed gases, and high heat and light generation. Unless these are controlled and directed properly, welders and those around them can be subjected to a variety of hazards resulting in injury and time lost from work. At the extreme, the potential for permanent injury and loss of life is very real.

Welding or Cutting Operation	Electrode Size: Metal Thickness or Welding Current	Filter Shade Number
Torch soldering		2
Torch brazing		3 or 4
Oxygen cutting		
Light	Under I in., 25 mm	3 or 4
Medium	1-6in.,25-ISOmm	4or5
Heavy	Over6in., 150mm	5or6
Gas welding		
Light	Under 1/8 in., 3 mm	4 or 5
Medium	1/8-1/2in.,3-l2mm	5or6
Heavy	Overl/2in., 12mm	6or8
Shielded metal arc	Under 5/32 in., 4 mm	10
welding (stick)	5/32-1/4 in., 4-6.4 mm	12
electrodes	Over 1/4 in., 6.4 mm	14
Gas metal arc welding (MIG)		
Ferrous base metal	All	11
Nonferrous base metal	All	12
Gastungsten arc welding (TIG)	All	12
Atomic hydrogen welding	All	12
Carbon arc welding	All	12
Plasma arc welding	All	12
Carbon arc air gouging		
Light		12
Heavy		14
Plasma arc cutting		
Light	Under 300 A	9
Medium	300to400A	12
Heavy	Over400A	14

TABLE 8-1 Recommended filter densities for welding, brazing, and cutting. (Courtesy American Welding Society)

CHAPTER 9

BLUEPRINT READING AND WELDING SYMBOL INTERPRETATION

Welding is a fabrication process. It is one of the final sequences after many other operations have been performed. The job sheet dictum which states, "weld to print," requires the welder to assemble, align, and otherwise arrange the various members into the geometric and dimensional model depicted on the blueprint.

Welders will find themselves functioning at various levels of responsibility. Simple production welding can involve only the tacking and welding of fixtured parts. In other instances the welder will be required to tack and weld precut and sized parts with the responsibility of maintaining squareness and any angular dimensions. At the highest level of responsibility, the welder/fabricator, working directly from the blueprint, will start from the rough cutting of stock to the construction of individual components and members, on to the fabrication of sub-assemblies and, finally, the assembly and welding of the complete unit.

As this journey is undertaken, the welder must "read" the print. In such reading it is well to remember that the blueprint is drawn by engineering and draft persons whose primary concern is the function of the part. How does it look? What does it do? Where does it fit? But for the welder/fabricator, the essence of blueprint reading is in interpreting the print as to how the part or unit is to be built. "Reading the print" then becomes the systematic and methodical interpretation of the various lines, symbols, views, notes, dimensions, and specifications. The welder must correlate all this information with the tools, equipment, and processes available in the shop. The welder/fabricator then plans a method and a sequence of operations to complete the job.

Blueprint Basics

Blueprints appear in two forms. The first is the assembly print (Fig. 9-1a). This drawing shows the entire weldment, which is often made up of several smaller components. The second type is the detail print (Fig. 9-1-b). Here, each part of the whole weldment is shown separately and usually enlarged for clarity and to allow complete dimensioning.

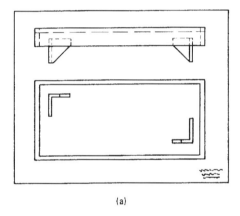

(a)

FIG. 9-1-a An assembly print showing all components.

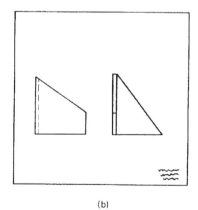

(b)

FIG. 9-1-b A detail drawing showing a single component of the whole assembly.

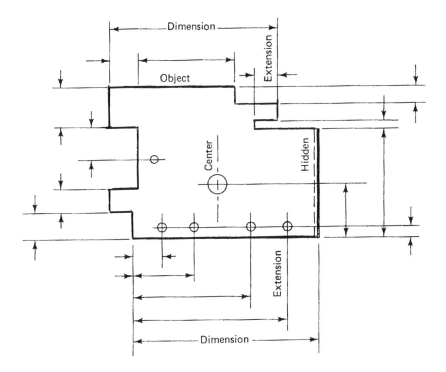

FIG. 9-2 The identification of various lines shown on a print.

All blueprints consist of several basic elements that perform specific functions.

♦ **Lines:** These are drawn light and heavy, solid, dashed, or broken. Lines show the outline of the object, aid in dimensioning, and form various symbols (Fig. 9-2).

♦ **Dimensions:** As whole numbers, fractions, decimals, or units of metric measure, dimensions show both the size of the part and indicate the locations of holes, cutouts, notches, and other details.

♦ **Notes and Specifications:** Notes are written instructions that cannot be shown by lines and symbols. Specifications are notes that give specific directions and information, such as welding processes, electrode types, or exact shop standards such as in finishing and machining. Today, notes and specifications are supplemented by "Geometric Tolerancing" symbols. Geometric Tolerancing specifies more than just a dimensional tolerance. The symbols indicate the function of the part or detail and, often replace written notes found on the print. For example, if a boss is to be welded to a flat surface, the perpendicularity of the axis of the boss in relation to the surface it is welded to is given a tolerance in addition to its dimensional location. Thus, as in figure 9-3, the perpendicularity of the boss must be within 0.015 inch to "Surface A" if the boss is to function properly after a hole is drilled through and tapped.

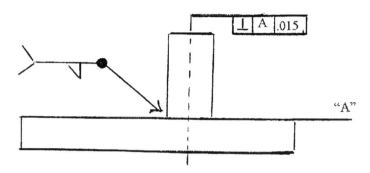

FIG. 9-3 Geometric tolerancing dimension and symbols for defining the function of the part.

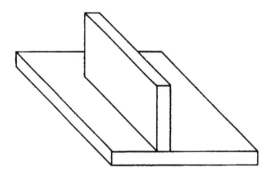

FIG. 9-4 A pictorial presentation presents difficulty in dimensioning and showing details.

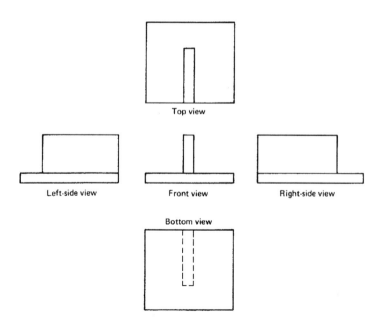

FIG. 9-5 An orthographic projection.

Blueprints usually provide views that show the object from several directions. The number of views depends on the complexity of the part and the amount of information that needs to be given to the fabricator. Blueprints are rarely, if ever, drawn pictorially or obliquely (Fig. 9-4). The pictorial presentation would cause problems in detailing hidden surfaces and edges. The blueprint is best drawn orthographically. As shown in figure 9-5, each side or "view" can be clearly detailed and dimensioned.

Blueprint Interpretation of Welded Items

Many drawings are produced by drafting departments in response to engineering requirements and they often contain explicit information and instructions pertaining to welding. This data may include aspects of joint preparation, root openings, welding sequence, and bead type and dimension, as well as pre and post-heat treatments. Instructions concerning the weld process to be used and the filler rod specifications are often included. In contrast are those drawings that show only the items required for general fabrication and contain only the note "WELD HERE." The welder/fabricator is then expected to use whatever judgment and expertise is called for to complete the job.

Working from a blueprint includes the following interpretations:

♦ What the drawing represents
♦ Type of material called for
♦ Size of the part(s) and scale of the print
♦ Number of parts to be made
♦ Use of stock sizes and shapes or the separate fabrication of
   components such as angles, channels, and round or square pipes
♦ Allowable deviation (tolerance) from dimensions given
♦ Method of joining
♦ Additional operations
♦ Finishing
♦ Selection of tools and machines
♦ Sequence of construction

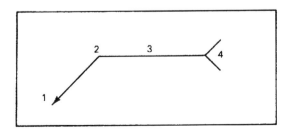

FIG. 9-6 The basic weld symbol format.

The blueprint may contain various welding symbols which are specific instructions to the welder. The basic welding symbol format and its component parts are shown in figure 9-6. The welding symbol is a form of shorthand. When placed strategically on a blueprint or sketch of a welded part, it conveys to the welder the information required to perform the welding operation according to the design specifications and to conform to accepted welding practices as well as the requirements of the job.

When properly drawn, and in turn, correctly interpreted, the welder can determine from the welding symbol:
- Exactly where the weld is to be located
- The amount of welding required
- The type of weld bead called for
- The size of the weld bead
- The type and amount of joint preparation
- Aspects of weld finishing
- The weld process to be used

The Composite Symbol

Figure 9-7 shows a composite symbol. Added to the basic format are individual symbol elements that indicate the basic joint type. Commonly used symbol elements are shown in figure 9-8. The complete AWS symbol chart (Fig. 9-16) can be found near the end of this chapter.

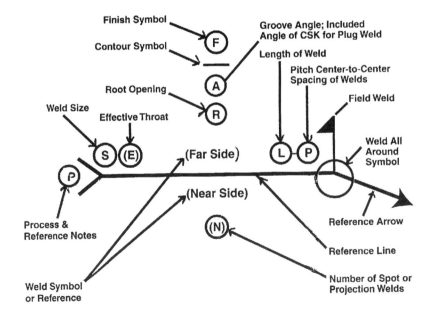

FIG. 9-7 The composite weld symbol.

FIG. 9-8 Symbol elements that depict the type of weld joint required.

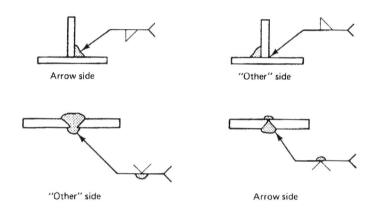

FIG. 9-9 The location of the symbol element above or below the reference line specifies the location of the weld, arrow side or other side.

Reading the Symbol

Referring back to figure 9-6:

♦ The arrow (1) points, as close as is practical, to the required location of the weld.
♦ The break (2) is the point on the symbol where certain information concerning field welds and the amount of welding is given.
♦ The location of a particular symbol element on the reference line (3) indicates on which side of the joint the weld is to be placed (Fig. 9-9).
   a. If the symbol is placed below the line, the weld is placed on the side the arrow points to.
   b. If the symbol is placed above the line, the weld is placed on the opposite or "other" side.
♦ Within the "V" of the tail (4) can be found the process specification (if not specified elsewhere on the blueprint) (Fig. 9-10).

FIG. 9-10 The weld process to be used is found in the tail of the symbol.

Some symbols and what they mean are shown in figures 9-11 through 9-15. The complete AWS welding symbol chart is shown in figure 9-16.

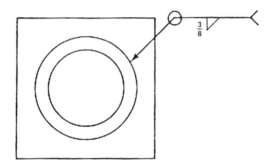

FIG. 9-11 The symbol for a continuous 3/8 bead all the way around the outside of the perimeter.

On the "other side" (inside perimeter)
weld a $\frac{1}{4}$ in. fillet weld all around

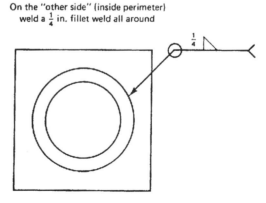

FIG. 9-12 The placement of the symbol element above the reference line call for the weld to be placed differently.

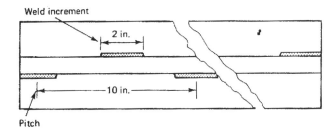

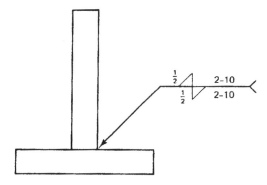

FIG. 9-13 This symbol calls for dimensioned weld increments staggered on both sides of the joint.

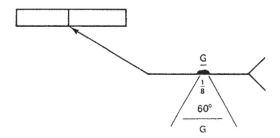

FIG. 9-14 These symbol instructions call for each plate to be beveled 30 degrees and separated by a 1/8 in. root opening. Joint is to be penetrated so that weld melt through is evident on the "other side." Both sides of the joint are to be ground flush.

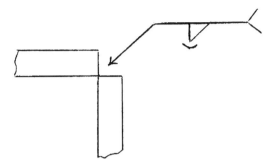

FIG. 9-15 The symbol specifies an outside corner fillet that is to be contour ground.

Basic Joint Types

All welding jobs use one or several of the basic joint types (Fig. 9-17 appears on page 60). In addition, the welder/fabricator may modify the joint to facilitate assembly and subsequent welding. The welder will often prepare the joint (joint preparation) to gain deeper penetration by grinding and/or flame cutting bevels and chamfers. This work is especially important if the welds must be ground flush or contoured to adjacent surfaces.

Joint-Type Selection

The decision to use a particular joint type and possibly to modify the joint in some way is based on consideration of one or more of the following:

♦ Overall shape of the weldment
♦ Effectiveness of the welding process used
♦ Accessibility of weld joint
♦ Amount of distortion that can be tolerated
♦ Amount of grinding and flame cutting required
♦ Amount of weld finishing required

An additional factor in joint-type selection is the actual strength required of the weld in relation to the type and amount of stress (load) expected. Types of loads are shown in figure 9-18, also on page 60. They are:

♦ a - tension
♦ b - compression
♦ c - shear
♦ d - bending
♦ e - torsion

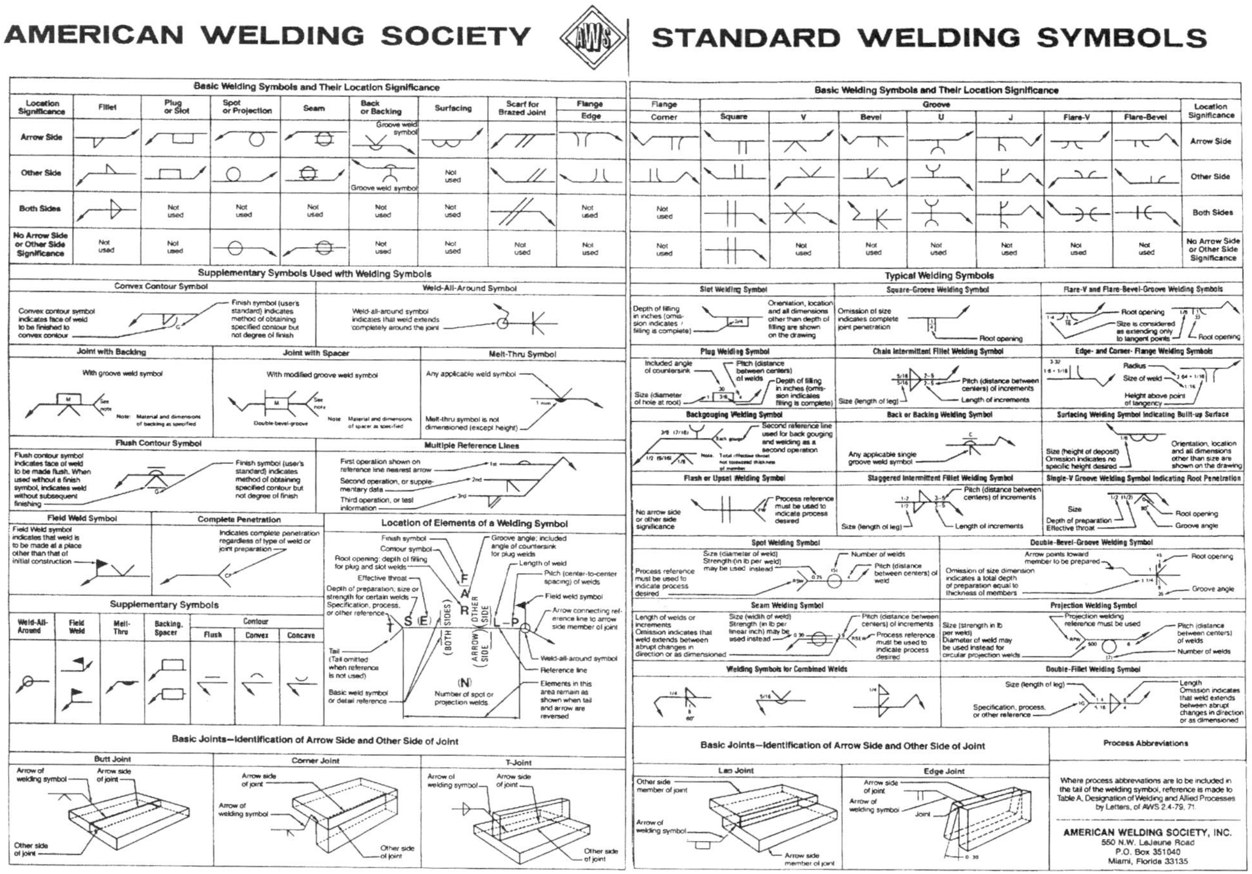

FIG. 9-16 Welding Symbol Chart (Courtesy of The American Welding Society)

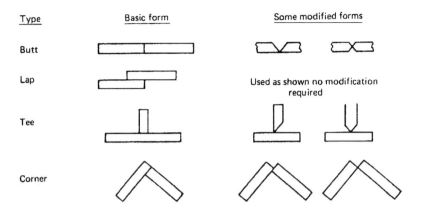

FIG. 9-17 Basic joint types for welding.

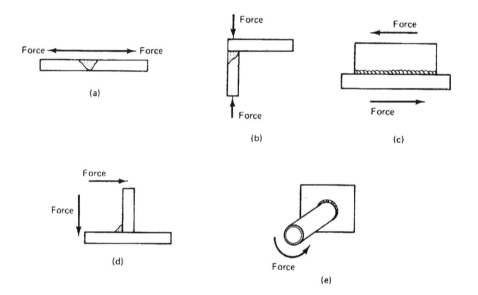

FIG. 9-18 Types of loads that can be exerted on a welded joint.

BASIC SHOP MATHEMATICS

The working welder, especially those engaged in shop fabrication, will require proficiency in basic computational mathematics. This means that the ability to quickly and accurately add, subtract, multiply, and divide whole numbers, fractions, and decimals will help greatly to ensure successful and continued employment. No matter how well parts are welded together, if nothing fits and components are misaligned due to errors in reading and calculating dimensions, those parts are of no use.

Tools

The welder must also be practiced in the use of various measuring tools Depending on the nature of the job, daily work situations call for the use of:

- ♦ Steel tapes and rules
- ♦ Combination and framing squares
- ♦ Protractors
- ♦ Micrometers
- ♦ Fractional and vernier calipers
- ♦ Height gauge and dial indicators

Skills

The specific mathematical skills required for welded fabrication include a working knowledge of:

- Geometric shapes, functions, and formulas
- Decimal equivalents of fractions
- Functions and measurements of angles
- Principles (at least) of trigonometry
- Basic algebra

The student should note that these subjects are included in any well-rounded high school curriculum

The welder/fabricator is expected to work from blueprints and sketches during the construction of whatever products are being manufactured. Together with the ability to operate and manipulate various welding and cutting equipment, a welder should be capable in the following related operations:

- Making of material lists based on a knowledge of stock sizes and shapes.
- Determination of actual cutting sizes, which often differ from any finished dimensions given on a blueprint.
- Calculation of weights based on square and linear measurements.
- Accurate reproduction of various geometric shapes.
- Calculations required in bending and forming operations using bend deduction and bend allowance formulas.
- Layout of holes and hole patterns in conformance with blueprint instructions.
- Selection of punch and die types and the determination of tonnage requirements in punching various thicknesses and alloys.

Most welding fabrication jobs will require the welder to convert the given "conventional" dimensions to "base line" dimensions (Fig. 10-1). So that

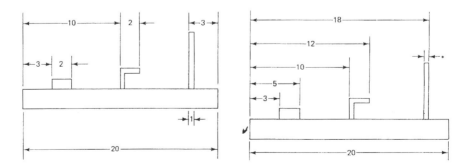

FIG. 10-1 Conventional dimensions are shown on the left, while the baseline method of dimensioning is shown on the right.

the weldment can be accurately assembled, the welder must be able to perform the basic mathematical computations involved. Additionally, the conversion of fractional dimensions to decimal equivalents is constantly being done as the job progresses.

Basic Rules of Fractions

The manipulation of fractional dimensions is all important if the job is to be completed satisfactorily. The following rules and examples for addition and subtraction can be used as a guide and review (Fig. 10-2).

EXAMPLE: Add ¼ + ⅜ + ³⁄₁₆.

$$\frac{1}{4} = \frac{4}{16}$$
$$\frac{3}{8} = \frac{6}{16}$$
$$+ \frac{3}{16} = \frac{3}{16}$$
$$\frac{13}{16}$$

EXAMPLE: Subtract ¹³⁄₃₂ from ⅞.

$$\frac{7}{8} = \frac{28}{32}$$
$$- \frac{13}{32} = \frac{13}{32}$$
$$\frac{15}{32}$$

FIG. 10-2 Adding and subtracting fractions.

Rule: *Fractions that are to be added or subtracted must have the same denominator. Before the addition and subtraction can be done, each fraction must be expressed in terms of a common denominator.*

When adding or subtracting mixed numbers, the same rule applies (Fig. 10-3).

EXAMPLE: Add 13⅞ and 6⁹⁄₁₆.

$$13\frac{7}{8} = 13\frac{14}{16}$$
$$+ 6\frac{9}{16} = 6\frac{9}{16}$$
$$19\frac{23}{16} = 20\frac{7}{16}$$

EXAMPLE: Subtract ¹³⁄₃₂ from ⅞.

$$\frac{7}{8} = \frac{28}{32}$$
$$- \frac{13}{32} = \frac{13}{32}$$
$$\frac{15}{32}$$

FIG. 10-3 Adding and subtracting mixed numbers.

Multiplication of Fractions (Fig. 10-4)

Rule: *When fractions are multiplied, the product of all the numerators is placed over the product of all denominators.*

EXAMPLE: Multiply ½ X ⅜ X ¼.

$$\frac{1}{2} \times \frac{3}{8} \times \frac{1}{4} = \frac{3}{64}$$

EXAMPLE: Multiply 5 X ⅜.

$$5 \times \frac{3}{8} = \frac{5}{1} \times \frac{3}{8} = \frac{15}{8} = 1\frac{7}{8}$$

FIG. 10-4 Multiplying fractions.

Division of Fractions (Fig. 10-5)

EXAMPLE: Divide ⅝ by ⁷⁄₁₆.

$$\frac{5}{8} \div \frac{7}{16} = \frac{5}{8} \times \frac{16}{7} = \frac{5}{\cancel{8}_1} \times \frac{\cancel{16}^2}{7} = \frac{10}{7} = 1\frac{3}{7}$$

FIG. 10-5 Dividing fractions.

Rule: *Invert the second fraction and multiply.*

Multiplication and Division of Mixed Numbers (Fig. 10-6)

Rule: *Convert the mixed numbers to improper fractions and apply the rules of multiplication and division.*

EXAMPLE: Multiply 3⅛ X 2½.

$$3\frac{1}{8} \times 2\frac{1}{2} = \frac{25}{8} \times \frac{5}{2} = \frac{125}{16} = 7\frac{13}{16}$$

EXAMPLE: Divide 4⅛ by 3½.

$$4\frac{1}{8} \div 3\frac{1}{2} = \frac{33}{8} \div \frac{7}{2} = \frac{33}{\cancel{8}_4} \times \frac{\cancel{2}^1}{7} = \frac{33}{28} = 1\frac{5}{28}$$

Note: Mixed numbers are converted to improper fractions by multiplying the whole-number part by the denominator, adding the product to the numerator, and placing this sum over the original denominator.

FIG. 10-6 Multiplying and dividing mixed numbers.

Fraction-to-Decimal Conversion

Common fractions are easily converted to decimals by dividing the numerator by the denominator (Fig. 10-7).

EXAMPLE: The decimal equivalent of ¼ is 0.250.

$$\begin{array}{r} .250 \\ 4\overline{)1.000} \\ \underline{8} \\ 20 \\ \underline{20} \\ 000 \end{array}$$

FIG. 10-7 Finding a decimal by division.

Decimal-to-Fraction Conversion

To convert a decimal to a fraction, approximately, multiply it by 64. The whole number portion of the resulting number will be the numerator of the fraction, and the denominator will be 64. Remember to reduce the fraction to its lowest terms (Fig. 10-8).

EXAMPLE: The fractional equivalent of .4375 is $\frac{7}{16}$.

$$.4375 \times 64 = 28$$
$$.4375 = \frac{28}{64} = \frac{7}{16}$$

EXAMPLE: The fractional equivalent of .610 is $\frac{39}{64}$.

$$.610 \times 64 = 39.04 \quad \text{(Round off to nearest 64th)}$$
$$.610 = \frac{39}{64}$$

FIG. 10-8 Finding a fraction by multiplication.

The ultimate goal of fabricators who frequently deal with fraction-decimal conversions is to memorize the commonly used equivalents and manipulate them without having to refer constantly to the conversion chart found in the appendix. Start by memorizing these basic conversions:

1 in.	= 1.000	1/16 in.	= 0.062
1/2 in.	= 0.500	1/32 in.	= 0.031
1/4 in.	= 0.250	1/64 in.	= 0.015
1/8 in.	= 0.125		

Note: *A three place decimal is accurate enough for general fabrication.*

Once this step is taken, manipulations of fraction-decimal conversions becomes a simple mental exercise (Fig. 10-9).

EXAMPLE: Find the decimal equivalent of $\frac{5}{32}$.

$$\frac{4}{32} = \frac{1}{8} = 0.125$$
$$\frac{1}{32} \quad\quad = 0.031$$

Therefore

$$\frac{5}{32} = 0.156$$

EXAMPLE: Find the decimal equivalent of $\frac{9}{64}$.

$$\frac{8}{64} = \frac{1}{8} = 0.125$$
$$\frac{1}{64} \quad\quad = 0.015$$

Therefore,

$$\frac{9}{64} = 0.140$$

FIG. 10-9 Once the basic conversions are memorized, other conversion are easily calculated.

Review of Basic Rules of Algebra
Addition

When combining numbers with like signs, add the numbers together and give the sum the same sign. When combining numbers with unlike signs, subtract the smaller from the larger and give the answer the sign of the larger number (Fig. 10-10).

EXAMPLES: $6 + 4 = 10$
$-6 - 4 = -10$

EXAMPLES: $10 - 12 = -2$
$-8 + 4 = -4$
$14 - 10 = 4$

FIG. 10-10 Combining like and unlike signs.

Multiplication and Division

When multiplying or dividing numbers with like signs, the answer is always positive. When multiplying or dividing numbers with unlike signs, the answer is always negative (Fig. 10-11).

EXAMPLES: $3(3) = 9$ 　　　 EXAMPLES: $(3)(-3) = -9$
$(-3)(-3) = 9$ 　　　　　　　 $\frac{3}{-3} = -1$
$\frac{-3}{-3} = 1$

FIG. 10-11 Multiplying and dividing numbers with like and unlike signs.

Algebraic Notations

Notation	Meaning
π	The Greek letter pi is a constant equal to 3.1416 (close approximation)
xy	x times y
$(x)(y)$	x times y
$14y$	14 times y
$3(x+y)$	$3x$ plus $3y$
$\dfrac{x}{y}$	x divided by y
x^2	x times x
x^3	x times x times x
x^{-2}	$\dfrac{1}{x^2}$
$x^2 x^3$	$x^{2+3} = x^5$
$\dfrac{x^4}{x^2}$	$x^{4-2} = x^2$
y^0	When the exponent is 0, the number is equal to 1

FIG. 10-12 The basic notations in algebraic expressions.

Transposing

The following rules of transposition must be followed when manipulating equations (Fig. 10-13 see page 68).

- Whatever is added or subtracted from one side of the equation must be added or subtracted from the other side.
- If one side of the equation is multiplied or divided by a number, the same operation must be done to the other side.

These basic examples can be applied to solving geometric equations. For example, if you have a piece of thin flat steel stock 38 inches long and you need to calculate the diameter of the circle that can be formed with this material, you must transpose the equation for circumference (Fig. 10-14 see following page).

Parentheses

If a portion of the equation is enclosed within parentheses, the operations inside the parentheses must be done first (Fig. 10-15).

EXAMPLES: Solve for x.

(1)
$$x + 10 = 12$$
$$x + 10 - 10 = 12 - 10$$
$$x = 2$$

(2)
$$x + y = a$$
$$x + y - y = a - y$$
$$x = a - y$$

(3)
$$3x = 18$$
$$\frac{{}^{1}\cancel{3}x}{\cancel{3}_{1}} = \frac{\cancel{18}^{6}}{\cancel{3}_{1}}$$
$$x = 6$$

(4)
$$\frac{x}{2} = 9$$
$$\frac{x}{\cancel{2}_{1}}{}^{1}(\cancel{2}) = 9(2)$$
$$x = 18$$

(5)
$$8x + 3 = 19$$
$$8x + 3 - 3 = 19 - 3$$
$$8x = 16$$
$$\frac{{}^{1}\cancel{8}x}{\cancel{8}_{1}} = \frac{\cancel{16}^{2}}{\cancel{8}_{1}}$$
$$x = 2$$

FIG. 10-13 Rules for transposing.

EXAMPLE:

$$c = \pi d$$ $c = 38$ inches
$\pi = 3.1416$

$$\pi d = c$$ Equation sides can be exchanged
without changing the value.

$$\frac{\pi d}{\pi} = \frac{c}{\pi}$$

$$d = \frac{c}{\pi}$$

$$d = \frac{38}{3.1416}$$

$$d = 12.7 \text{ inches}$$

FIG. 10-14 Transposing terms in an equation.

EXAMPLES: $(8 + 10)(2 + 3) =$

$(18)(5) = 90$

$86 - (8 + 2 + 6) =$

$86 - 16 = 70$

FIG. 10-15 Working with parentheses.

Terms

Those portions of an equation separated by either plus (+) or minus (-) signs are called terms. In the equation 18 + 8 - 2(6), 18, 8, and -2(6) are the terms. Any multiplication or division operation within a term must be done before terms can be combined (Fig. 10-16).

EXAMPLE: $18 + 8 - 2(6) =$
$18 + 8 - 12 = 14$

FIG. 10-16 Working with terms.

Geometric and Trigonometric Formulas

The following geometric equations and trigonometric functions will help in performing many typical fabrication assignments. Anything that is fabricated, whether it is a small precision sheet metal part or a large structural assembly, will conform to one or more of the basic geometric shapes. Aside from the obvious use of geometry and trigonometry to derive indirect measurements, the fabricator must use these formulas to determine material requirements. Calculating the amount of material needed will also allow you to determine material weight and cost.

The Circle

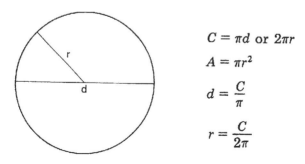

$C = \pi d \text{ or } 2\pi r$

$A = \pi r^2$

$d = \dfrac{C}{\pi}$

$r = \dfrac{C}{2\pi}$

where C = circumference
A = area
r = radius
d = diameter
π = 3.1416

FIG. 10-17 The circle.

Properties of Geometric Shapes
The Hollow circle

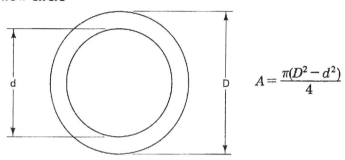

$$A = \frac{\pi(D^2 - d^2)}{4}$$

FIG. 10-18 The hollow or inner circle.

Segments of a Circle

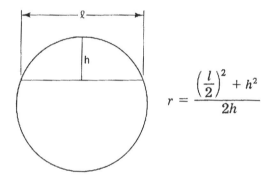

$$r = \frac{\left(\frac{l}{2}\right)^2 + h^2}{2h}$$

FIG. 10-19 A circle segment.

Rectangle

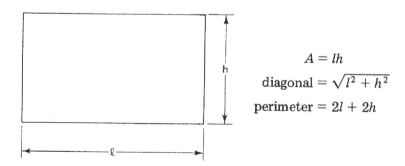

$$A = lh$$
$$\text{diagonal} = \sqrt{l^2 + h^2}$$
$$\text{perimeter} = 2l + 2h$$

FIG. 10-20 The rectangle.

Sector of a Circle

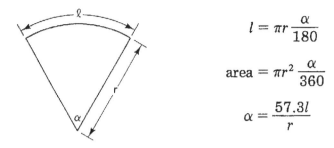

$$l = \pi r \frac{\alpha}{180}$$

$$\text{area} = \pi r^2 \frac{\alpha}{360}$$

$$\alpha = \frac{57.3l}{r}$$

FIG. 10-21 A circle sector

Circular Cone

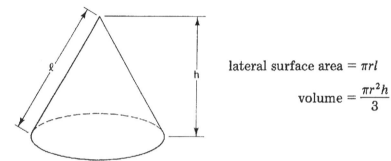

$$\text{lateral surface area} = \pi r l$$

$$\text{volume} = \frac{\pi r^2 h}{3}$$

where r denotes the base radius.

FIG. 10-22 The cone.

Frustrum of a Cone

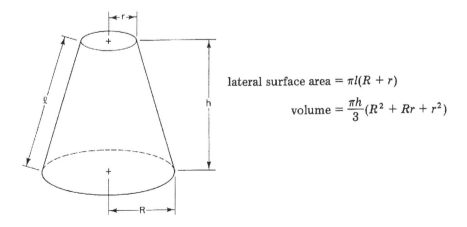

$$\text{lateral surface area} = \pi l (R + r)$$

$$\text{volume} = \frac{\pi h}{3}(R^2 + Rr + r^2)$$

FIG. 10-23 The frustrum of a cone.

Cylinder

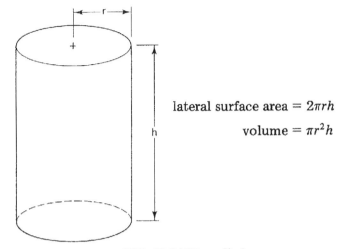

lateral surface area $= 2\pi rh$

volume $= \pi r^2 h$

FIG. 10-24 The cylinder.

Cube

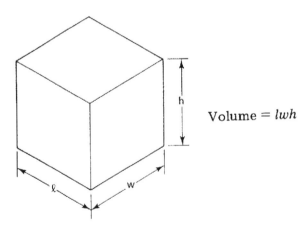

Volume $= lwh$

FIG. 10-25 The cube.

Sphere

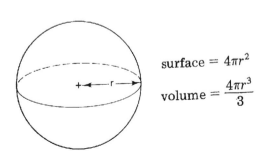

surface $= 4\pi r^2$

volume $= \dfrac{4\pi r^3}{3}$

FIG. 10-26 The sphere.

Trapezoid

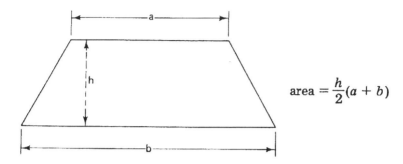

$$\text{area} = \frac{h}{2}(a + b)$$

FIG. 10-27 The trapezoid.

Calculating the capacity of a cylindrical tank is done by using the formula for cylinder volume and applying the following conversions (Fig. 10-28):

$$\text{volume} = \pi r^2 h$$

$$= 3.14(5^2)(30)$$

$$= 3.14(25)(30)$$

$$= 2355 \text{ cubic inches}$$

$$\text{Capacity in gallons} = \frac{2355}{231} = 10.19 \text{ gallons}$$

FIG. 10-28 Formula conversions for calculating capacity.

Cylinder

For example, how many gallons will a tank hold that measures10 inches inside diameter and 30 inches long? (see Fig. 10-29 on next page.)

231 cubic inches will hold 1 gallon.
1 cubic foot will hold 7½ gallons.
1728 cubic inches = 1 cubic foot.
31½ gallons = 1 barrel.

$$\text{volume} = \pi r^2 h$$

or

$$\text{volume} = 0.7854 d^2 h$$

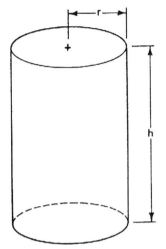

FIG. 10-29 Calculating the capacity of a cylinder of a given size.

Complementary and Supplementary Angles

Complementary angles total 90 degrees. If given angle is 30 degrees, it's compliment must be 60 degrees. Supplementary angles total 180 degrees. If a given angle is 25 degrees, it's supplement is 155 degrees.

Right Triangles

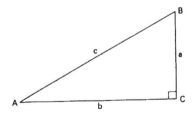

Trigonometric function	Example
Sine of an angle $= \dfrac{\text{opposite side}}{\text{hypotenuse}}$	$\sin A = \dfrac{\text{side } a}{\text{side } c}$
Cosine of an angle $= \dfrac{\text{adjacent side}}{\text{hypotenuse}}$	$\cos A = \dfrac{\text{side } b}{\text{side } c}$
Tangent of an angle $= \dfrac{\text{opposite side}}{\text{adjacent side}}$	$\tan A = \dfrac{\text{side } a}{\text{side } b}$
Cotangent of an angle $= \dfrac{\text{adjacent side}}{\text{opposite side}}$	$\cot A = \dfrac{\text{side } b}{\text{side } a}$

FIG. 10-30 The right triangle.

Formulas for Solving Right Triangles

Formulas for Solving Right Triangles

Measurements given	a	b	c	A	B
			Measurement to find		
a, b			$a^2 + b^2$	$\tan A = \dfrac{a}{b}$	$\tan B = \dfrac{b}{a}$
a, c		$c^2 - a^2$		$\sin A = \dfrac{a}{c}$	$\cos B = \dfrac{a}{c}$
b, c	$c^2 - b^2$			$\cos A = \dfrac{b}{c}$	$\sin B = \dfrac{b}{c}$
A, a		$a \cot A$	$\dfrac{a}{\sin A}$		$90 - A$
A, b	$b \tan A$		$\dfrac{b}{\cos A}$		$90 - A$
A, c	$c \sin A$	$c \cos A$			$90 - A$
B, a		$a \tan B$	$\dfrac{a}{\cos B}$	$90 - B$	
B, b	$b \cot B$		$\dfrac{b}{\sin B}$	$90 - B$	
B, c	$c \cos B$	$c \sin B$		$90 - B$	

FIG. 10-31 solving right triangles.

Oblique Triangles (Fig. 10-32)

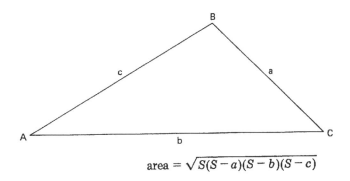

$$\text{area} = \sqrt{S(S-a)(S-b)(S-c)}$$

The value for S is calculated as follows:

$$S = \frac{a + b + c}{2}$$

**Solutions for Oblique Triangles When Two Sides
and an Included Angle Are Known**

$$a^2 = b^2 + c^2 - 2bc \cos A$$
$$b^2 = a^2 + c^2 - 2ac \cos B$$
$$c^2 = a^2 + b^2 - 2ab \cos C$$

FIG. 10-32 Oblique triangles.

Metric System (Linear)

The metric measurement system is based on powers of 10, with the meter being the fundamental unit. The system is easy to work with because all units are related and different values can be expressed by placing and moving the decimal point rather than by juggling fractions.

Beginning with the meter and to express values in smaller units, **divide by 10**, moving the decimal point to the left.

1 meter = 39.37 in.
0.1 meter = 3.937 in. (a decimeter)
0.01 meter = 0.3937 in. (a centimeter)
0.001 meter = 0.03937 in. (a millimeter)

To express higher values, **multiply by 10**, moving the decimal point to the right.

1 meter = 39.37 in.
10 meters = 393.7 in. (a dekameter)
100 meters = 3937. in. (a hectometer)
1000 meters = 39370. in. (a kilometer)
Note: The terms dekameter and hectometer are not generally used.

And, of course 39370 inches equals 3,280 feet which in turn is approximately 0.6 mile. The difficulty in adapting shop work to the metric system is in constantly having to compare one type of measurement to the other. When the global economy dictates the universal use of metrics, all blueprints and the calibration of the machinery to manufacture the parts will be in metrics and this will eliminate the comparison problem. A complete metric conversion table can be found in the appendix.

C H A P T E R

WELDING METALLURGY

By simple definition, metallurgy is the study of metals. As a field of study, however, it includes a range of knowledge and expertise well beyond the requirements of practical shop and field welding. It has been shown that the success of welding as a method of joining, is due primarily to both the effectiveness of the various welding processes and the skill of the welder. But when we go beyond actual application of welding skill into the areas of serviceability and endurance, it becomes necessary to think of the weldment in metallurgical terms.

Through variations in chemical composition and various heat treating processes during manufacture, metals derive certain mechanical properties. These properties include:

1. Hardness
2. Brittleness
3. Ductility
4. Toughness
5. Grain size
6. Thermal conductivity
7. Thermal expansion
8. Response to heat treatment
9. Tensile strength

The glossary in Section III will provide exact definitions of these terms

Effects of Chemical Variation and Welding Heat

The weld itself and areas adjacent to the weld should be considered to be remanufactured metal. The welder should therefore be at least familiar with the effects of potential chemical variations and with the effects of welding heat. The two basic effects of heat are tempering, or degree of hardness, and annealing, which means loss of temper or induced softness. Which effect becomes significant in any welding situation depends on the presence and amount of chemical and alloying elements in the metal being joined. A partial list of the principal elements and their effects is as follows:

1. Carbon: increases hardness
2. Silicon: improves response to heat treatment
3. Manganese: increases hardness and tensile strength
4. Sulfur: increases machinability but produces porosity
5. Chrome: increases corrosion resistance
6. Nickel: increases toughness
7. Molybdenum: increases hardness and toughness

Metal Identification

The initial task of the welder is the identification of the metal to be welded. Accurate identification is required before the proper procedures, rod selections, welding technique, and heat treatment can be determined. Metal identification methods include:

1. Spark pattern test (grinding) for steel alloys (Fig. 11-1)
2. Magnetic (ferrous or nonferrous) and some stainless steels
3. Chip (hardness)
4. Fracture (grain structure)

It is often difficult, especially with aluminum and magnesium castings, to determine the material. Take some filings from the casting, place them on a small piece of paper and ignite the paper. Magnesium filings will flare brightly when the flame reaches them.

Additional Factors to Consider

Characteristics of particular types of steel, such as stainless and the high-strength alloys, together with the nonferrous metals such as aluminum and magnesium, must be considered carefully in welding. These metals, and certain others, must be able to maintain degrees of corrosion resistance and strength not only in the weld area, but in the adjacent heat-affected zone as well. Most welded fabrications are made of the low carbon or mild steels, the alloy steels,

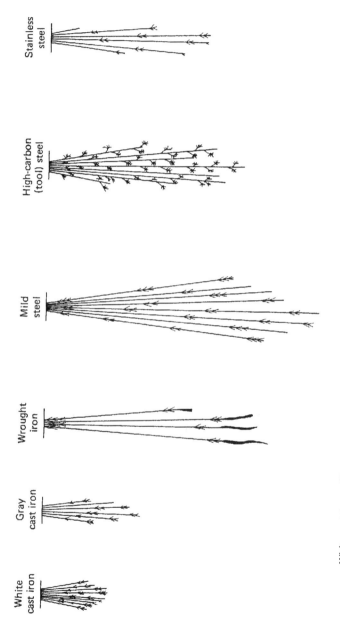

White cast iron: Short, small spark pattern, red at the start and straw yellow at the end.

Gray cast iron: Short, very small spark pattern, red at the start and straw yellow at the end.

Wrought iron: Long spark pattern, few sparks, straw yellow at the start and white at the end.

Mild steel: Long spark pattern, moderate amount of sparks, white in color.

High-carbon tool steel: Moderate pattern length, many sparks, white in color.

Stainless steel: Moderate pattern length, few sparks, straw yellow at the start and white at the end.

FIG. 11-1 The spark pattern of known samples is compared to those of material to be identified.

the stainless steels, and various aluminum alloys. Discussion of approaches to the welding of these metals follows.

Mild Steels

Most fabricated and welded products are made of low carbon steel, either hot or cold rolled. They would typically bear the SAE designation 1018 through 1024. With carbon contents indicated as being in the 1/4 percent range and lower, these steels form and weld very well with no pre or post-heat treatment required. The decision to use either the hot rolled or cold rolled variety would depend on the welding method. As the hot rolled steel surface is heavily oxidized in the rolling process, flakes of these oxides tend to contaminate the tungsten electrode used in the GTAW process. Other than degreasing, cold rolled steel requires no other preparation and is better suited to that process. Both GMAW and SMAW are better suited for welding the hot rolled material, as this type is less costly and more available in a wider variety of shapes and sizes.

Low carbon steels are also of the "rimmed" or "killed" variety. These terms refer to the residual oxygen left in the steel, with "killed steel" having less oxygen. If porosity is a problem when welding either type, use filler rods with higher amounts of deoxidizers such as titanium, silicon, and aluminum. When purchasing steel plates and sheets for welded fabrication, those designated as "universal mill" will be the most economical but may cause fabrication and welding problems as the content analysis is unreliable. For best quality and to meet any code requirements, steels that bear ASTM (American Society for Testing Materials) or Federal QQ-A designations are of guaranteed analysis.

Alloy Steels

Many steels used in construction, ship building, and in construction equipment manufacture are of the low alloy-high strength variety and are often known by their proprietary trade names. The common defect of these steels in respect to welding lies in their susceptibility to hydrogen induced cracking and in the need to replace elements liberated from the joint during welding. As a result of these effects, filler rod selection and pre- and post-heat treatments are critical. SMAW welding dictates the use of those electrodes known as the low-hydrogen types. GMAW, even as a low-hydrogen process, still requires the use of filler wire with specific analysis and content.

Stainless Steel

Stainless steel is an alloy that contains significant amounts of chromium. The material is highly corrosion resistant and extremely strong. Depending on alloy type, the average tensile strength will be approximately 85,000 psi.

There are four basic types of stainless steel, bearing the American Iron and Steel Institute (AISI) numerical designations 410, 430, 302, and 202. The primary alloy content of each is as follows:

Type   Content

410   Chromium, 12 percent
430   Chromium, 17 percent
302   Chromium, 18 percent; nickel, 8 percent
202   Chromium, 18 percent; nickel, 5 percent; manganese, 8 percent

Additional alloy types are derived by adding specific elements in closely controlled amounts. These additions will change or enhance a particular alloy's corrosive resistance, strength, or most important from the fabricator's standpoint, the material's workability. Stainless steels are broadly classified as being hardenable by various methods according to their basic grain structure. The austenitic type, which includes both the 200 and 300 series numbers, are hardened only by cold-working operations. The martensitic type, which includes most but not all of the 400 series, are hardenable either by air or furnace cooling or by quenching. The ferritic type includes a group of 400 designations that is not hardenable.

Most stainless types are nonmagnetic, but several alloys in the 400 group are magnetic in all conditions. Additionally, alloys have been developed with particular properties that enhance high-temperature service. These fall into a 500 series designation or may have special manufacturer's designations.

For welded fabrication, which includes resistance welding and brazing as well as extensive bending and forming, type 304 is recommended. All the stainless alloys have good welding and machining qualities, but careful review of the manufacturer's specifications will reveal which is best in each instance. Obviously, for welding, the hardening types will require an additional post-heat treatment operation. Careful consideration must also be given to the end use application. Stainless steel is used in the medical, chemical, and food-processing industries, so resistance to particular corrosive environments is an important criterion for alloy selection as is the use of the correct filler rod.

The two primary difficulties encountered when welding stainless steel are distortion and carbide precipitation. Distortion in stainless steel is caused by its tendency to heat up quickly as welding current passes through it. This is because stainless is less conductive than ordinary mild steels. Welding of stainless will require 20 to 30 percent less amperage than similar cross sections of mild steel.

Carbide precipitation is also due to the tendency of stainless to overheat during welding. The condition occurs when in essence the chromium in the fused weld area precipitates or comes "out of solution" and resolidifies outside the weld bead, leaving the weld non-stainless. Such conditions are caused by multipass welding, large beads, and the juncture of two or more beads. To help control the problem, copper chill bars, skip welding, and other techniques that control distortion and help prevent the weldment from getting too hot can be employed. When called upon to weld and fabricate stainless steel, it is always advisable to closely follow the supplier's recommendations.

Aluminum

Aluminum and its various alloys are commonly welded with either the GTAW or GMAW processes. Before development and refinement of those methods, aluminum was arc welded (SMAW) or gas welded (OAC). Stick welding could not routinely produce dense, X-ray quality joints, and gas welding required a highly corrosive flux that became active only 100 or 200 degrees below aluminum's melting temperature of around 1150°F.

The aluminum sheets, plates, and shapes that are commonly fabricated by welding are approximately 96 percent pure, with the balance being the alloying elements needed for a particular application. The series designations for these basic types, and the primary alloying elements, are:

1. 1000 series - commercially pure (99 percent)
2. 2000 series - copper
3. 3000 series - manganese
4. 4000 series - silicon
5. 5000 series - magnesium
6. 6000 series - silicon and magnesium
7. 7000 series - zinc

Within these series designations are approximately 150 separate alloy types, which meet a variety of end-use demands and workability criteria. The following paragraphs describe the characteristics and workability of those alloys most commonly fabricated in the metal shop. In addition to the basic numerical designation there is often a letter suffix that denotes a degree of hardness has to be induced by either cold working (H) or heat treatment (T). The letter "O" indicates the softest type within an alloy group. Melting temperatures vary slightly with alloy content.

The following alloys are the most commonly encountered in shop fabrication and welding:

♦ 2024-T3 - cannot be fusion welded or brazed, leaving resistance
   spot welding and riveting as the only practical joining methods
♦ 5052-H32 - highly weldable by fusion methods and spot welding
   but does not braze well. Other alloys in the 5000 series such as 5083
   and 5086 are classified as the "marine grades" due to their even higher
   magnesium content.

- ◆ 6063-T5 - known as the "architectural grade" this alloy is fabricated as extruded shapes (angles, channels, etc.) into furniture, cabinetry, and a variety of building components. This alloy is heat treatable and welding may produce a loss of strength in the immediate weld area. Fusion welding is most applicable, and the alloy brazes very well.
- ◆ 6061-T6 - this alloy is used for most of the structural applications of aluminum. It is highly fusion weldable, brazes well and can also be spot welded. As a heat treatable alloy, the "T-6" temper is completely lost in the fused weld area and in the heat affected zone (HAZ) (Fig. 11-2).

To achieve full hardness and strength throughout such weldments, it is required that the full thermal treatment be repeated. This treatment includes, at specified temperatures and time duration, annealing (stress relief), solution heat treatment (quenching), and artificial aging (holding-oven time). If this procedure is not possible or practical, the loss of strength due to welding can be compensated for by increased cross-section dimensions and other design factors. Without such post-heat treatment, T6 aluminum will return naturally to a T4 condition in the fused area after about 48 hours.

Series 6061 alloy is also used for fabrication and joining by the dip brazing process. Purchased in the O temper, intricate assemblies and shapes can be formed and held together by tack welding (using #1100 filler rod) or by employing slots and tabs. The joints to be brazed are painted with a flux, then the entire assembly is immersed in a bath of molten braze material. The braze flows through the prepared joints and at the same time the alloy is solution heat treated to the T4 condition.

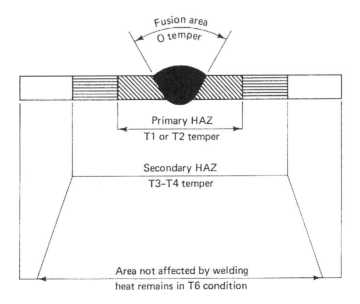

FIG. 11-2 The heat affected zone (HAZ) extends beyond the fusion area.

Aluminum weldments, particularly in sheet metal, are prone to cracking. Wherever possible, joints should be fully penetrated, with weld reinforcement appearing on the opposite surface. Weld dressing should leave rounded or radiused internal and external corners. Weld jigs and fixtures should allow for part shrinkage. Upon solidification molten aluminum will shrink up to 10 percent by volume.

Magnesium

The welder today is most likely to see magnesium in the form of a casting that may require repair or modification. It is the lightest of all structural metals, with aluminum being 1-1/2 times heavier. Thus for a time it was in direct competition with aluminum for use in air frame construction. But as a fabricated metal in either sheet or bar stock, its workability has proven to be too conditional in most fabrication and welding applications. Magnesium parts can be welded and brazed. Fusion welding is best accomplished with the inert-gas processes.

Solid magnesium stock will not easily ignite and burn; however, finely divided particles such as grindings, chips, and turnings are a definite fire hazard. Accumulation of such debris must be disposed of promptly and properly.

Brass, Copper, and Other Metals

These alloys can usually be fusion welded with great success. However, due to their variety and often proprietary characteristics, it is always best to follow the manufacturers recommendation very closely. The same recommendations apply to titanium, which is finding more popular use as a fabricated metal. Welding titanium can be very difficult, requiring the use of the inert gas processes. Often the welding is done within an enclosed atmosphere of inert gas. High strength forgings and castings are welded with the electron beam process, which overcomes titanium's adverse reaction to welding heat and atmospheric corrosion at elevated temperatures.

CHAPTER 12

DISTORTION AND THE HEAT EFFECTS OF WELDING

Welding has been defined as the "re-manufacturing" of the base metals. When virgin metal is manufactured, certain heat treatment procedures and cycles are employed to impart specific mechanical properties, some of which were discussed in Chapter 11. Temperature changes caused by the heat of welding can have two negative effects on the weldment. One is the changing of the basic metallurgical structure of the metal and the second is in the dimensional and angular geometry of the workpiece.

Not only is the heat of welding a factor, but when welding is preformed in an environment susceptible to varying amounts of atmospheric contamination, critical metallurgical changes can occur. These contaminants are primarily the elements oxygen, nitrogen, and hydrogen, which will combine with the molten weld metal to form new compounds that can be detrimental to the integrity of the welded joint. Yet, as will be shown in Chapter 16, when combined with certain inert gases these elemental gases can actually improve weld quality and process efficiency. From a more immediate and practical standpoint, the problem of weld distortion is one that can cause the greatest discouragement, profit loss, and dissatisfaction with welding in general.

Distortion (Fig. 12-1) is a result of the expansion and, upon cooling, the shrinkage of the material being welded. If

FIG. 12-1 The addition of a weld bead will distort an unrestrained plate.

a piece of plain metal material is subjected to a heating and cooling cycle while lying on a work surface, there is little noticeable effect as the material has not been restrained, restricted, or gained the addition of a weld bead of any kind. Moreover, if the heating and cooling was timed, controlled, and evenly applied as in an oven, distortion is relatively slight or even non-existent. In contrast, even a simple weldment will react to heating and cooling in adverse ways. The more complicated the joint geometry and extra additions of weld bead material, the more potential for warping and distortion. This potential will vary not only with the thickness and bulk of the weldment but also with the type of metal being welded. Because the thermal conductivity of stainless steel is only about one-third that of mild steel and its coefficient of thermal expansion is one and one half times that of mild steel, distortion is more of a problem in welding the stainless steels.

When comparing the potential for distortion in aluminum in relation to mild steel, there are some interesting contradictions. The coefficient of expansion is about twice that of steel so the shrinkage effect upon cooling is higher. Thus aluminum is more prone to cracking during cooling, but aluminum's higher thermal conductivity (approximately four times that of steel) transfers heat to adjacent plates much quicker. The resulting lower temperature differential between the weld and the HAZ should result in less distortion, especially in thinner weldments of relatively simple design. Yet, because many aluminum weldments employ thicker cross sections to overcome the alloy's lower strength, distortion can be as problematic as when welding mild steel.

Weld shrinkage and distortion can never be entirely eliminated but it can be effectively controlled by careful and intelligent procedures as suggested in the following list.

♦ Avoid over welding — large and unnecessary multiple beads will increase shrinkage. Employ proper edge preparation and joint fit up techniques.

- Use intermittent welds—more is not always better and will always produce more distortion.
- Use as few weld passes as possible —weld with the largest diameter electrodes possible.
- Place welds as close as possible to the neutral axis —to provide the least leverage to pull the plate members out of alignment.
- Balance weld placement —make shrinkage forces work against each other.
- Back step the weld bead—if the overall direction of the weld is left to right, deposit each segment of the bead in right to left increments.
- Anticipate shrinkage —pre-position or pre-stress parts so that the normal shrinkage will move the parts to where they should be. Another method is to clamp identical weldments back to back, allowing them to cool before removing the restraints.
- Plan a logical welding sequence —plan the placement of each successive bead so that the shrinkage of each will cancel out the effect of the other.
- Neutralize shrinkage after welding —peening, either by hand or mechanically, will counteract the shrinkage as the weld cools. The procedure must be done judiciously, usually being employed between multiple passes. Most codes prohibit peening root and cap passes. Post stress relieving of the whole weldment will remove this accumulated shrinkage force, and is done ideally in a specially designed oven with heating, holding, and cooling time controls.

These basic methods are illustrated in figure 12-2. The geometry of the part will vary, but it is always advisable to complete the welding in the shortest possible time and to make every effort to minimize the unequal buildup of heat in the adjacent and adjoining members. Careful pre-heating using temperature indicating marking crayons, and further steps to retard cooling if only by burying the work in an insulating blanket, will be effective in controlling distortion.

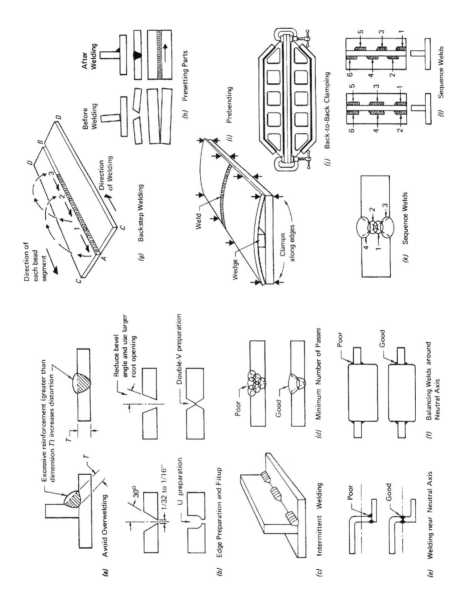

FIG. 12-2 Various methods of controlling distortion. (Courtesy of Lincoln Electric Company)

CHAPTER 13

OXY-ACETYLENE WELDING, BRAZING, AND CUTTING

Process and Equipment

The oxy-acetylene welding process utilizes the heat of a gas flame, consisting of oxygen and acetylene, to melt the surface of the base metal and produce a fusion weld. Filler rod must be added to obtain proper bead contour (Fig. 13-1). Acetylene, combined with oxygen, produces the hottest gas flame obtainable (6300°). Alternative fuel gases (Mapp, propane, etc.) are available but produce lower flame temperatures and are limited in use to heating and flame cutting.

The acetylene cylinder is a low-pressure steel tank filled with a porous material. This porous material is saturated with acetone, which can absorb large amounts of acetylene and stabilize it for safe storage. Full cylinder pressure is about 250 psi at 70°F. Four safety fuse plugs are incorporated into the tank and designed to melt if the temperature of the cylinder rises above a safe level. Because of the liquid acetone content, the tank must always be used in the upright position and the valve should never be opened more than one full turn.

The oxygen cylinder is a one-piece, seamless, steel, high-pressure tank. Its valve assembly has two important fea-

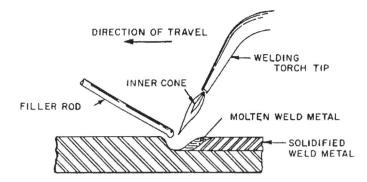

FIG. 13-1 The gas welding process. (Courtesy Hobart Institute of Welding Technology)

tures, the first of which is a two-way seat. When the valve is fully closed, it seals the cylinder. When fully open as it seals the stem to prevent leakage. During cylinder operation, the valve should be fully opened. The assembly is also equipped with a safety disk which is designed to rupture and release the compressed oxygen if the cylinder pressure exceeds a safe level. A full cylinder will be pressurized at about 2200 psi at 70°F.

Oxygen and acetylene regulators are designed to reduce cylinder pressure and maintain that pressure without constant readjustment. Reduction is accomplished by the spring-loaded valve attached to the flexible diaphragm (Fig. 13-2). When the adjusting screw is turned in (clockwise), the compression spring forces the valve assembly to open. As gas is consumed and the cylinder pressure begins to drop, the compression spring will force the diaphragm valve assembly to open further to maintain constant working pressure. This balance between compression spring and the gas pressure in the cylinder maintains a uniform gas flow to the torch. The regulators are equipped with two gauges which indicate cylinder and working pressures respectively.

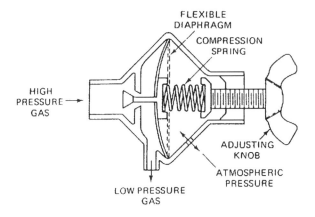

FIG. 13-2 A typical gas regulator.

The twin hose is actually two separate hoses joined by a web. The acetylene (red) hose has left-hand thread fittings, and the oxygen (green) hose has right-hand thread fittings, to prevent accidental interchanging of the hoses. All acetylene hose fittings have a groove cut in them to indicate a left-hand thread.

The welding torch or blowpipe consists of an oxygen valve, acetylene valve, body mixing chamber, and welding tip (Fig. 13-3). Its function is to mix the acetylene and oxygen together and facilitate adjustment to the desired flame. A good quality torch will be constructed of brass with the exception of the tip, which is copper. Tip size is determined by the orifice or opening in the end of the tip. The larger the opening, the greater the flame size.

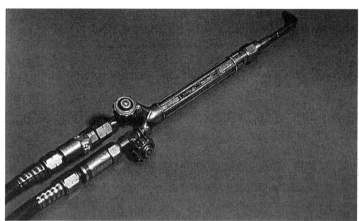

NOTE: Tip size numbering varies from one manufacturer to the next. Each manufacturer will provide, in catalog or pamphlet form, a recommended tip size and gas pressure chart for your particular torch.

FIG. 13-3 The gas welding torch will accept a variety of tip sizes.

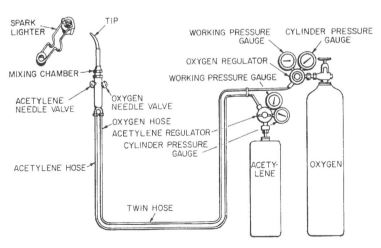

FIG. 13-4 Components for gas welding and cutting. (Courtesy Hobart Institute of Welding Technology)

Equipment Setup

- Chain the oxygen and acetylene cylinders securely in the cart.
- Open the oxygen cylinder valve slightly to remove any dirt or dust. A quick opening and closing of the valve is sufficient.
- Do the same with the acetylene cylinder using a T-handle wrench. The T-handle must remain on the cylinder so that it can be turned off quickly if a hose fire occurs.
- Attach the oxygen regulator to the oxygen cylinder and the acetylene regulator to the acetylene cylinder. Tighten both securely. Check to make sure that the adjusting screw on each regulator is loose (regulator closed).
- Fasten the twin hose to the regulators, red left-hand thread on acetylene, green right-hand thread on oxygen.
- Secure the twin hose to the welding torch following the same left and right-hand directions.

Pressurizing and Lighting

- Stand to one side and open the oxygen cylinder valve slowly . Rapid opening may cause regulator damage or rupture. After the cylinder pressure has moved up, open the valve all the way.
- Using the T-handle wrench, slowly open the acetylene cylinder. When the cylinder pressure gauge starts to move, open the valve no more than one full turn.
- Crack open the acetylene valve on the torch and turn in the regulator adjusting screw until the desired working pressure is reached. Shut the torch valve.
- Crack open the oxygen valve on the torch and turn in the regulator adjusting screw until the desired working pressure is reached. Shut the torch valve.
- At this time, check all fittings and connections for leaks using a soapy water solution. Fix all leaks before lighting the torch.
- Point the torch away from you, open the acetylene valve, and light the torch using a spark lighter. After the torch is lit, open the oxygen valve and adjust the flame.

Shut Down
- Shut off the torch by closing the acetylene valve first and then the oxygen valve. Oxygen does not burn but merely supports combustion, so that it is actually the acetylene that is flammable. Following the above procedure will insure that the flame is extinguished at the tip and not inside the torch or possibly in the hose.

♦ Shut off the acetylene cylinder valve.
♦ Open the acetylene torch valve and wait until both gauges on the acetylene regulator read zero.
♦ Back out the acetylene regulator adjusting screw until it becomes loose. Then shut off the acetylene torch valve.
♦ Shut off the oxygen cylinder valve.
♦ Open the oxygen torch valve and wait until both gauges on the oxygen regulator read zero.
♦ Back out the oxygen regulator adjusting screw until it becomes loose. Then shut off the oxygen torch valve
♦ Roll up the hoses and secure the equipment.

Safety

♦ Safety check valves should be installed on all oxy-acetylene out fits. These valves prevent the reverse flow of one gas into the other line.
♦ While in use, keep cylinders vertical, especially the acetylene.
♦ Always use the proper welding goggles. Never use ordinary sunglasses.

Adjusting the Torch

Most difficulties with OAW can be attributed to improper flame adjustment. There are three flame adjustments.

♦ Neutral: The neutral flame (Fig. 13-5a) burns equal amounts of acetylene and oxygen and provides maximum welding temperatures. It is used for fusion welding mild steel and flame cutting. Because the neutral flame burns equal amounts of each gas, it will not alter the chemical composition of the molten weld metal by carbon additions.

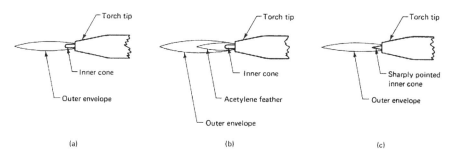

FIG. 13-5 a. The neutral flame for fusion welding. b. The carburizing flame for silver brazing and when brazing aluminum. c. An oxidizing flame for brazing with bronze alloy filler rods.

♦ Carburizing: The carburizing flame burns with an excess amount of acetylene and varies in temperature depending on the length of the acetylene feather (Fig. 13-5b). This flame is used primarily for silver brazing and welding aluminum.
♦ Oxidizing: The oxidizing flame burns with an excess amount of oxygen and can reach a temperature of 6300oF (Fig. 13-5c). It is used for braze welding cast iron with a bronze filler rod. This flame should never be used for fusion welding of mild steel because it will cause excessive porosity and weld brittleness.

Torch Adjustment

♦ Set the oxygen and acetylene regulator working pressures to coincide with the welding tip size being used.
♦ With welding goggles on, open the torch acetylene valve one full turn and ignite the torch using a spark lighter.
♦ Slowly open the torch oxygen valve. As oxygen is added to the flame, the inner cone and acetylene feather will appear. This is the carburizing flame.
♦ Gradually increase the oxygen until the acetylene feather and inner cone come together. At this point the flame is neutral, and this is the adjustment used for all fusion welding of mild steel.
♦ The addition of more oxygen will cause the inner cone to become pointed and a hissing sound may develop. The flame is now oxidizing.
♦ Experiment with the various flame adjustments until they can be performed without difficulty.
♦ When shutting the torch down, close the acetylene valve first and then the oxygen.

Operating pressures, as dictated by the tip size, are set primarily at the regulators. The torch valves are for minor flame adjustments (controlling the acetylene feather) and to turn off the torch. Using the torch valves to overly restrict the flow of gas may produce another source of operating difficulty .

Running a Weld Bead

Note: **Although OAW finds limited use in production welding, the practice it affords will help the beginner become familiar with fusion concepts and help develop the manual dexterity needed for more advanced processes.**

1. Take several pieces of 1/16 by 3by 6-inch hot-rolled sheet steel and wire brush to remove loose scale and rust from the surfaces.

2. Select several RG45 1/16-inch diameter welding rods.
3. Install the recommended welding tip for 1/16-inch steel (each manufacturer will specify a tip number for a given thickness).

To begin practice:

1. Light the torch and adjust to a neutral flame.
2. Place one of the steel pieces on a firebrick.
3. Hold the torch at a 45-degree angle in the direction of travel and transversely perpendicular (Fig. 13-6). The end of the inner cone should be 1/16 to 1/8 inch off the surface of the plate.
4. Make small circular motions with the torch (Fig. 13-7). When a puddle of approximately 3/16 inch diameter forms, move across the plate, being careful to maintain a uniform bead width.
5. Repeat Step 4 until a uniform bead can be made easily.
6. To add filler rod, form a puddle and dip the end of the filler rod into it (Fig. 13-8)

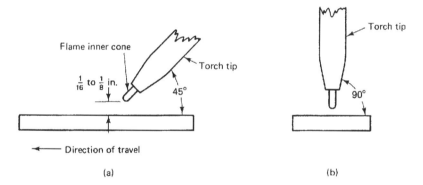

FIG. 13-6 The innner cone of the gas flame should never actually touch the work.

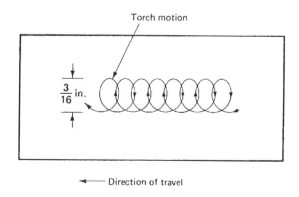

FIG. 13-7 Create a molten puddle before adding filler rod.

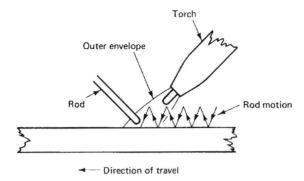

FIG. 13-8 Rod is added to the puddle's leading edge.

7. As you progress across the plate, continue to dip the rod repeatedly and uniformly. The amount of filler rod added will determine the bead crown.

NOTE: Always keep the end of the filler rod within the outer envelope of the flame.

Forehand Welding Technique

The forehand method of oxy-acetylene welding produces a bead of shallow to medium penetration. This technique is generally used for welding thin-gauge metal or when higher welding speed is required. Begin practice by joining two pieces of plate with a simple butt weld.

To begin forehand gas welding:

1. Place two of the steel pieces together on a firebrick, edge to edge, with a root opening of approximately 1/16 inch.
2. Light the torch and adjust to a neutral flame.
3. Tack-weld the pieces together (Fig. 13-9) to ensure proper alignment.
4. Hold the torch at a 45-degree angle in the direction of travel and transversely perpendicular.

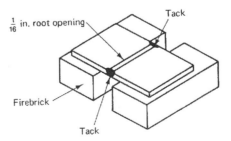

FIG. 13-9 Preparing two pieces of 1/16 in. thick steel for practice.

5. Begin just inside one of the tacks. Make small circular motions with the torch and when the edges of each plate melt (puddle should be approximately 1/4 inch diameter), add filler rod and begin moving in the direction of travel.
6. Sufficient filler rod must be added to fill the root opening and produce the desired bead crown.
7. Add filler rod with a steady dipping motion.
8. Fill the crater at the end of the bead.

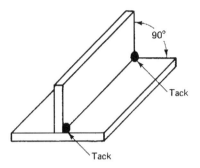

FIG. 13-10 Tacking a "tee" joint.

To practice a fillet weld:

1. Secure two of the steel pieces into the T-joint position and tack-weld (Fig. 13-10).
2. Hold the torch at a 45 -degree angle in the direction of travel and at a 45-degree angle transversely (Fig. 13-11).
3. Making small circular motions with the torch, form a puddle on each of the plates and begin to add filler rod. Initially, the torch should be directed more toward the bottom plate because it will not puddle as easily as the top.
4. Add filler rod continuously along the top edge of the puddle (Fig. 13-12). This will help prevent undercutting.
5. The face of the bead should be flat and approximately 1/4 inch to 3/8 in. wide.

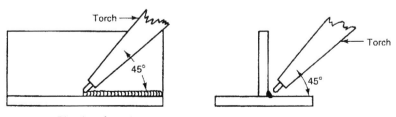

FIG. 13-11 Welding the "tee" joint with the proper torch angle.

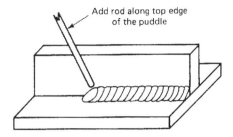

FIG. 13-12 Rod is added to the topedge of the puddle.

Backhand Welding Technique

The backhand oxy-acetylene welding technique produces a bead with medium to deep penetration. It is generally used for welding of heavier-gauge plate and, pipe, and for beads requiring maximum penetration.

1. Take two pieces of 1/4-inch by 2-inch by 6-in. hot-rolled steel and make a 35-degree bevel on one edge of each.
2. Take two pieces of 3-inch schedule 40 pipe, 4 inch long, and bevel one end of each to a 35-degree angle.
3. Remove all loose scale and rust with a wire brush.
4. Select several RG45 1/8-inch-diameter welding rods.
5. Install the recommended welding tip for 1/4-inch steel.

The procedure for backhand gas welding is as follows:
1. Place the steel plates together, edge to edge, on a firebrick with a root opening of 3/32 inch.
2. Adjust a neutral flame and tack-weld the plates at each end.
3. Hold the torch at a 60-degree angle away from the direction of travel and transversely perpendicular (Fig. 13-13).
4. Moving the torch in a crescent motion, form a puddle on each plate edge, and add the filler rod using the dip method (Fig. 13-14). Add enough rod to produce the desired bead crown.
5. Move slowly enough to ensure maximum penetration.

The following practice is for small diameter black iron pipe:
1. Clamp the pipe edges together with a root opening of 1/16 inch
2. Adjust to a neutral flame and tack-weld the pipes together at three evenly-spaced points.
3. Hold the torch at a 45-degree angle away from the direction of weld and transversely perpendicular.
4. Moving the torch in a crescent motion, form a puddle on each pipe edge and add filler rod using the dip method. Add enough rod to produce the desired bead crown.
5. Move slowly enough to ensure maximum penetration. Learn to "read" the puddle. The width of the puddle indicates the amount of penetration.

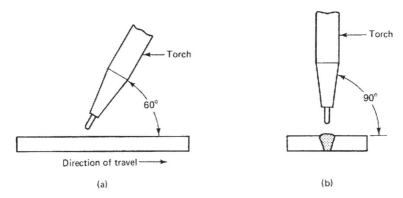

FIG. 13-13 A back hand welding technique is used on thicker materials.

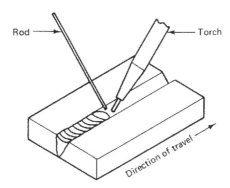

FIG. 13-14 Rod is added in greater amounts to fill the joint and assure some penetration if needed.

Brazing Mild Steel

Brazing is a process by which metals are joined by capillary action with a nonferrous filler rod having a melting temperature above 800oF but less than that of the base metal. The base metal is not melted or fused. Braze welding occurs at higher, but still not at the base metal melting temperature. There is some surface alloying, increasing the strength of the joint.

The most common brazing filler rod is bronze, which has a melting temperature of 1600°F. The rod consists mainly of copper and zinc with small amounts of silicon, tin, iron, and manganese. These additional elements increase the free-flowing action of the rod, decrease its tendency to fume, and help deoxidize the base metal surface.

Flux: Brazing flux is made of borax and boric acid. It is a white granular powder and is usually packaged in 1-pound cans. The function of the flux is to remove oxides from the surface of the base metal, allowing the bronze filler rod to flow smoothly and evenly over the entire joint area. The flux may be applied by one of the following methods:

- The end of the filler rod is heated and dipped into the flux can. The flux will adhere to the heated area.
- The powdered flux may be mixed with hot water to form a paste. The paste can then be brushed on to the braze area.
- The powdered flux can be sprinkled on the heated joint.

The basic steps for brazing include:
1. The base metal must be cleaned and degreased to remove any loose scale, rust, paint, oil, or grease.
2. Light the torch and adjust to a slightly oxidizing flame.
3. Heat the end of the filler rod and dip it into the flux can. The flux will adhere to the heated area.
4. Hold the torch at the angle shown in Fig. 13-15 and heat the base metal until a dull red spot approximately 3/8 inch in diameter appears.
5. Add filler rod with a dipping motion and progress along the joint maintaining a constant bead width. The diameter of the red spot will determine the width of the bead.
6. When the fluxed portion of the filler rod is consumed, repeat Step 3.
7. Proper technique will produce a slightly crowned, smoothly flowing, evenly rippled bead. If the plate is too cold, a globular, poorly fused, uneven bead will result. Overheating will cause an overly large bead, burned plate, and excessive fuming, as indicated by the white powdery appearance of bead edges.

Overheating of the base metal will also cause the bronze filler rod to fume. The fume smoke can be toxic; therefore, brazing should be done in a well-ventilated area.

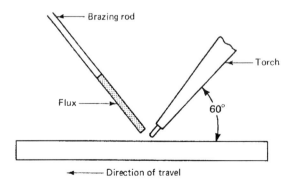

FIG. 13-15 Torch angle used for brazing.

Butt and Fillet Braze Welds

Joint design for brazing is basically the same as that for fusion welding. For butt welds made on plates thicker than 1/8 inch, the plate edges should be filed or ground to form a V-groove to ensure maximum penetration. Metal surfaces in and around the weld area should be thoroughly cleaned and degreased. Heavy rust or scale should be ground off. All bench welds should be done on firebrick to avoid the heat loss caused by direct contact with a metal-top table.

The procedure for a butt braze is as follows:
1. Position the plates on pieces of firebrick and set a root opening of slightly less than 1/16 inch.
2. Tack the joint at each end to ensure proper lineup.
3. Hold the torch at a 30-degree angle in the direction of the weld and transversely perpendicular.
4. Heat each side of the weld joint to tinning temperature (dull red) and add filler rod. Check for complete tinning and penetration (Fig. 13-16).
5. Take care to maintain a uniform bead width and equal distribution of braze rod on each plate.
6. If more than one bead is required, be sure that the filler metal being deposited is completely fused to the previous bead.

The procedure for a fillet braze is:
1. Position the plates at a 90-degree angle to each other on a fire brick.
2. Tack the joint ends to maintain proper lineup.
3. Hold the torch as shown in figure 13-17.
4. Heat the weld joint until each plate is a dull red.
5. Add filler rod and maintain a uniform bead width.
6. Care must be taken to ensure that the filler material flow completely into the root of the joint.

FIG. 13-16 Brazing a butt joint.

Brazing Cast Iron

Begin by thoroughly cleaning the weld area to remove oil, grease, and paint, and grind away all surface scale and rust from the weld joint and surrounding area. Grind a bevel on all broken edges. Care should be taken not to bevel completely through the edge, leaving part of the broken section intact to aid in alignment. See "Cast Iron" in SMAW Chapter. If the casting is only cracked, drill small-diameter stop holes at each end of the crack, which will prevent the crack from traveling during the welding operation. V-groove the crack edges with a grinder.

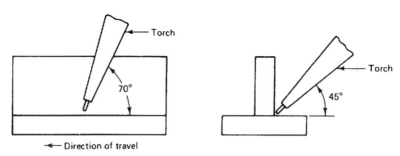

FIG. 13-17 Brazing a "tee" joint.

Use the following procedure:

1. Preheat the casting to a temperature of 500 to 900°F. This heaing will help prevent warping and reduce the chance of additional cracking in highly stressed areas.
2. A borax-boric acid flux will give good results; however, flux containing brass spelter will greatly aid in tinning the cast iron.
3. Heat the weld joint to a dull red and add filler rod (bronze). Hold the torch at the same angle indicated for brazing mild steel. Use a moderately oxidizing flame.
4. Make sure that tinning is uniform and complete.
5. If more than one pass is required, each pass must be completely fused to the one that preceded it.
6. The completed weld should be slightly wider than the V-groove.
7. The casting should be allowed to cool slowly by either burying the work in sand or by placing the repair in an oven.

Silver Brazing

Silver brazing is a process by which metals are bonded together with a silver-alloy, nonferrous filler rod, having a melting temperature of 1600° to 2200°F depending upon the silver content. When the brazing joint is heated to tinning temperature, the filler metal is added and drawn into it by capillary action. The silver brazing filler rod consists primarily of silver with percentages of copper, zinc, and nickel. Its most important properties are corrosion resistance, high electrical conductivity, and the ability to withstand constant vibration.

The most common silver brazing flux used is a fluorine compound packaged in paste form. Application is made by painting the flux paste onto the weld joint surfaces. This type of flux can be irritating to the skin and may give off toxic fumes. It should be used in a well ventilated area. Brazing joints must be close fitting so that the bond can be made with a minimum amount of filler metal. Best results in silver brazing are obtained when the parts are separated by no more than 0.005 in. The strength of the joint will decrease as the part separation increases. The type of joint best suited for silver brazing is the lap (Fig. 13-18).

Common Oxy-acetylene Weld Defects and Causes

Defect/Problem	Cause
Backfire	1. Overheated tip 2. Loose or dirty tip 3. Improper flame adjustment 4. Incorrect gas pressures
Flashback	1. Welding tip orifice clogged 2. Incorrect gas pressures
Poor fusion	1. Welding tip too small 2. Base metal not properly melted before adding filler rod
Bead cracks	1. Weld bead not proportional to the size of the base metal 2. Excessive heat 3. Rapid cooling
Brittle weld beads	1. Incorrect flame adjustment 2. Wrong filler rod
Poor penetration	1. Faulty joint preparation 2. Welding speed too fast 3. Wrong torch angle
Excessive penetration	1. Welding tip too large 2. Weld speed too low
Excessive warping and distortion	1. Weld bead not proportional to the base of metal thickness (overwelding)
Globular bead appearance (brazing)	1. Weld joint was not heated to proper tinning temperature 2. Dirty weld joint
Uneven tinning (brazing)	1. Dirty joint 2. Flux not properly applied 3. Uneven heat distribution
Excessive bead width (brazing)	1. Poor heat control; too large an area was heated to tinning temperature
Excessive fuming and white powder on brazed weld joints	1. The base metal was overheated

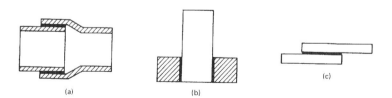

FIG. 13-18 Typical silver brazing joints.

To set up a practice silver braze joint:

♦ Practice on degreased cold rolled steel, stainless steel, or brass..
♦ Paint a thin layer of paste flux on the joint surfaces.
♦ Light the torch and adjust to a carburizing flame. The acetylene feather should be approximately 1 in. long.
♦ Heat the joint uniformly to tinning temperature. The brushed-on flux will have a watery appearance as the tinning temperature is approached. When tinning temperature is reached, the rod will melt and the metal will be drawn into the joint. Care must be taken to avoid overheating the joint.
♦ Use just enough filler metal to fill the joint. No outside reinforcement is required.
♦ After the brazing is completed, excess flux can be removed with hot water, or by soaking in a chemical solvent made expressly for this purpose.

The Flame Cutting Process and Equipment

Flame cutting is a process in which oxygen reacts with heated steel to cause rapid oxidation (burning). A spot on the steel is heated to its ignition temperature (cherry red). A stream of high-pressure oxygen is then introduced and a thin section of metal is burned away. The narrow gap that results from the burning is called the kerf (Fig. 13-19).

The two types of torches used most frequently are the one-piece manual cutting torch and the cutting attachment, which is mounted on a regular welding torch body (Fig. 13-20). The cutting torch differs from the welding torch in that it has a second oxygen passage and control lever for the high-pressure oxygen used to burn the metal. Cutting nozzles are made of copper and consist of a circle of preheat orifices and a cutting oxygen orifice, and are designed to serve two functions. The first is to preheat the steel to ignition temperature (approximately 1600°F) and the second is to direct high-pressure oxygen on the heated area to cause rapid oxidation and that will burn away the metal.

In general, nozzle size, oxygen pressure, and kerf width will increase as the thickness of the steel being cut increases. Table 13-1 can be used as a guide in determining these variables. Because of differences in manufacturer's nozzle numbering systems, nozzle sizes on the chart are listed by preheat and cutting orifice diameters.

Straight-Line Cutting

To begin, clean the surface of the plate to be cut to remove any paint, excessive rust, or scale. With a soapstone, mark guidelines on the plate and position the plate so that the line of cut clears the edge of the welding table.

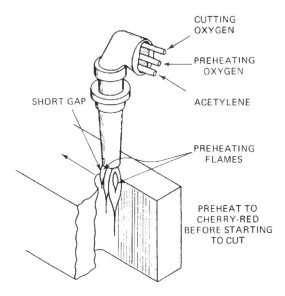

FIG. 13-19 The oxy-acetylene flame cutting process.

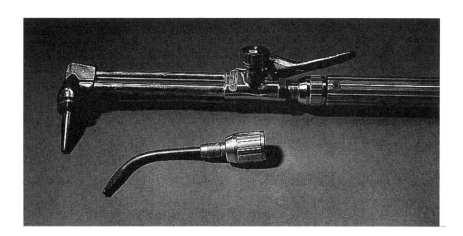

FIG. 13-20 The gas welding torch can be fitted with a cutting attachment.

| Metal Thickness (in.) | Nozzle Size | | Acetylene (psi) | Oxygen (psi) |
	Preheat Orifice Drill Size	Cutting Orifice Drill Size		
1/4	71	64	5	30
3/8	69	57	5	35
1/2	69	57	5	40
3/4	68	55	5	45
1	68	55	5	50
2	66	53	5	55
4	63	47	5	60

TABLE 13-1 Gas cutting tip sizes.

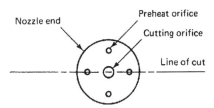

FIG. 13-21 Careful alignment of the pre-heating orifices will help cutting efficiency.

Install the proper cutting nozzle and set the recommended gas pressures. If the nozzle has four preheat orifices, see figure 13-21 for proper alignment.

To perform basic straight line cutting:

1. Both hands must be used in order to obtain a high-quality cut. Hold the torch with one hand gripping the cutting oxygen lever. Make a fist with the other hand and rest the torch on it.
2. Before lighting the torch, move it along the line of cut to make sure that the cutting line can be easily followed.
3. Light the torch and adjust a neutral flame. Depress the cutting oxygen lever to check the adjustment; the flame should remain neutral.

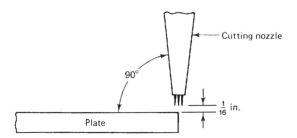

FIG. 13-22 As in welding, the inner cones should be above the work.

4. Hold the torch so that the cutting nozzle is at the edge of the plate and perpendicular to it (Fig. 13-22). The ends of the preheat flame cones should be about 1/16 in. above the plate surface.
5. Heat the edge of the plate to ignition temperature (cherry red) and depress the cutting oxygen lever all the way.
6. Move the torch along the line of cut at a slow, steady rate. If the torch is moved too quickly, the preheat will be lost and the cutting will stop. If this happens, return to the point at which the cut stopped and repeat Step 5.
7. Remove all slag from the completed cut.
8. A straight edge guide may be used if a more accurate cut is required (Fig. 13-23).

FIG. 13-23 Using a piece of angle for a straight edge guide.

Beveling and Piercing Technique

The preparation for bevelling and piercing is the same as for straight-line cutting, with the following exception: If a four-preheat-orifice cutting nozzle is used, see Fig. 13-24 for the proper alignment for bevel cutting.

A bevel cut can best be made with the use of an angle-iron guide (Fig. 13-25). First position the cutting nozzle against the angle-iron guide so that the preheat flame closest to the plate is approximately 1/16 in. above the surface, then heat the plate to ignition temperature and squeeze the cutting oxygen level. The cutting speed must be reduced on a bevel cut. Finally, remove all slag from the completed cut.

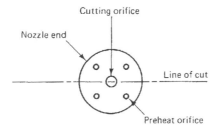

FIG. 13-24 By slightly altering the orientation of the preheating orifices, bevel cutting is made easier.

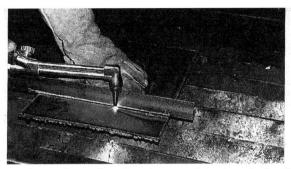

FIG. 13-25 Inverting a piece of angle sets a consistent angle for the torch tip.

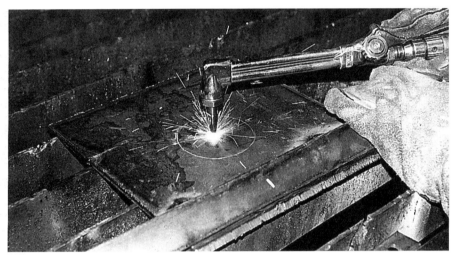

FIG. 13-26 Hole piercing can clog the cutting tip if not done properly.

To pierce a plate:

♦ Hold the lighted torch perpendicular to the plate with the preheat flame cones approximately 1/16 in. from the surface (Fig. 13-26).
♦ Heat the plate to ignition temperature and slowly squeeze the cutting oxygen lever. Raise the torch slightly to prevent slag from fouling the nozzle.
♦ After the initial hole is pierced in the plate, lower the torch so that the preheat flame cones are again 1/16 in. above the plate surface and complete the cutting of the desired hole diameter.
♦ Remove all slag from the completed cut.

Pattern Cutting

The easiest and most accurate method of flame cutting a circle is by using a circle-cutting guide (Fig. 13-27). Such a guide may be purchased commercially made in the shop.

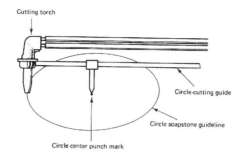

FIG. 13-27 Using a circle cutting attachment.

◆ Strike a heavy center punch mark at the center of the circle
◆ Adjust the cutting guide to the radius of the circle.
◆ Light and adjust a neutral flame, position the torch and cutting guide on the plate, pierce a hole through the plate, and cut the circle.
◆ Remove slag from the completed cut.

For straight-line patterns:

◆ Draw the pattern on the plate surface with soapstone.
◆ Strike a center punch mark at each corner of the pattern. This will aid in cut alignment if the soapstone guidelines are burned off.
◆ Cut each straight-line segment freehand or with a straightedge guide.
◆ Remove all slag from the completed cut.

For curved line patterns:

◆ Draw the pattern on the plate surface with soapstone.
◆ Strike a center punch mark every 1/2 inch along the pattern line (Fig. 13-28). This will make the pattern line more visible during the cutting operation.
◆ Cut the plate using the freehand technique.
◆ Remove all slag from the completed cut.

The flame cutting procedures described here can be performed mechanically with specifically designed equipment. Straight line and bevel cutting can be

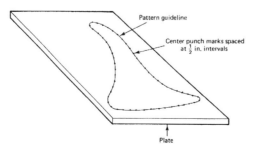

FIG. 13-28 Marking a pattern cut with a center punch.

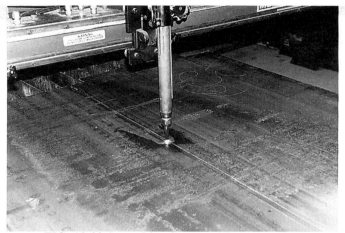

FIG. 13-29 A machine flame cutting operation.

mechanized as shown in figure 13-29. Pantograph flame cutters, some with multiple torches, can reproduce intricate shapes with almost machine-cut quality edges.

Safety:

- ◆ Be aware of the increased fire and burn hazard of flame cutting. The area directly under the line of cut should be kept clear.
- ◆ The welder must pay particular attention to the position of the twin hose, as it can be cut or punctured if a hot drop-off falls on it. Always keep the hoses behind you, away from the cutting area.
- ◆ Proper protective equipment must be worn to prevent burns caused by flying sparks and molten metal.
- ◆ If the cut is to be made in a direction other than the floor, make sure that the area in which the sparks are landing is clear of personnel and flammable material.

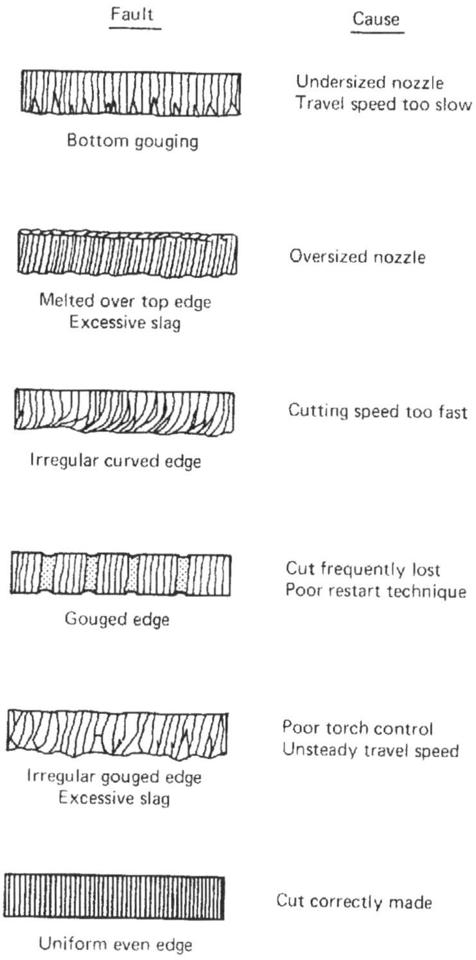

Fault	Cause

Bottom gouging

Undersized nozzle
Travel speed too slow

Melted over top edge
Excessive slag

Oversized nozzle

Irregular curved edge

Cutting speed too fast

Gouged edge

Cut frequently lost
Poor restart technique

Irregular gouged edge
Excessive slag

Poor torch control
Unsteady travel speed

Uniform even edge

Cut correctly made

FIG. 13-30 Flame cutting problems and remedies.

Common Flame-Cutting Defects and Causes

Defect/Problem	Cause
Poor appearance	1. See Fig. 13-30
Preheat cannot be maintained	1. Cutting nozzle too small 2. Preheat flames too short 3. Cutting speed too fast
Kerf closes in and fuses together	1. Cutting nozzle much too large 2. Preheat flames too long 3. Cutting speed too slow
Cutting torch backfires	1. Clean or replace the defective nozzle
Excessive slag buildup on bottom edge of cut	1. Dirty cutting nozzle 2. Oxygen pressure too high 3. Unsteady cutting speed 4. Preheat flames too long
Molten metal is propelled upward (spits back)	1. Improper preheat 2. Cutting nozzle too small 3. Oxygen pressure too low

CHAPTER 14

Shielded Metal Arc Welding

Process and Equipment

Shielded metal arc welding (SMAW) is a fusion process using an electric arc as a heat source and a consumable electrode as a filler. The arc is formed between the tip of the electrode and the surface of the metal being welded (Fig. 14-1). The intense heat generated melts the surface of the metal and the electrode, and intermixes the two, solidifies, and forms a weld with the same metallurgical and strength characteristics as the parent metal. The heat of the arc also transforms the electrode coating into a gaseous shield. A slag is formed over the weld bead which is removed with a chipping hammer and wire brush.

Power Supply Selection

The transformer-type power supply converts line voltage into electrical power suitable for welding by reducing (stepping down) the voltage and increasing the amperage. Transformers will output only AC welding current and are usually preferred for shop use because of their quiet operation, low maintenance, and low initial cost.

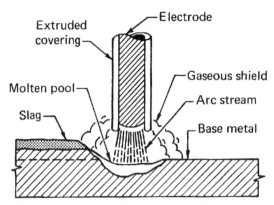

FIG. 14-1 The shielded metal arc welding process. (Courtesy Hobart Institute of Welding Technology)

A more productive version of the AC transformer is the AC/DC rectifier. A rectifier is an electrical device used to convert alternating current to direct current. This conversion enables the transformer/rectifier welding machine to produce both AC or DC welding current, and allows the welder greater process and electrode flexibility.

The inverter power supply is a "state of the art," DC welding machine. Input line voltage can be either single or three phase, allowing the inverter to be used for AC or DC arc welding and the TIG/MIG processes along with their control systems. Inverters are extremely light weight and portable. A 300 amp inverter can weigh less than 90 lb.

Welding generators are designed primarily to produce DC current. They are driven by an electric motor or engine (gasoline or diesel). Most modern generators are of the dual control type, which allows the welder greater flexibility in controlling the volt-ampere characteristic. In other words, the welder has greater control over the intensity of the arc. When thin material is welded, a soft, shallow penetrating arc is desirable, whereas thicker material requires a more forceful arc for maximum penetration. Engine-driven generators are used primarily for field welding. These machines are self-contained and require no external power. Many are also capable of producing auxiliary power for lights and electric tools.

Power Supply Equipment

Specifically designed for the welding power supply, welding cables are constructed of fine strands of copper wire wrapped in a rubber insulator. Cable size is determined by the amperage output of the power supply and cable lead length (TABLE 14-1). Cable connections must be made with either silver brazed or mechanical-type cable lugs.

Weld Type	Welding Current (A)	Length of Welding Cabk Circuit (ft)					
		50	100	150	200	300	400
Manualor	75	6	6	4	3	2	1
semiautomatic	100	4	4	3	2	1	1/0
welding	150	3	3	2	1	2/0	3/0
(up to 60%duty cycle)	200	2	2	1	1/0	3/0	4/0
	250	2	2	1/0	2/0	4/0	—
	300	1	1	2/0	3/0	—	—
	350	1/0	1/0	3/0	4/0	—	—
	400	1/0	2/0	3/0	—	—	—
	450	2/0	3/0	4/0	—	—	—
	500	3/0	3/0	4/0	—	—	—
Semiautomatic or	400	4/0	4/0	—	—	—	—
automatic welding	800	2-4/0	2-4/0	—	—	—	—
(60 to100% duty cycle)	1200	3-4/0	3-4/0	—	—	—	—
	1600	4-4/0	4-4/0	—	—	—	—

TABLE 14-1 Suitable cable lengths for SMAW.

The electrode holder is designed to carry the welding current to the electrode while holding it firmly in place. Although various types of holders are available, the most common is the spring jaw, fully insulated, grip type (Fig. 14-2). The size of the holder is determined by the power supply output. A ground clamp secures the ground cable lead to the workpiece. The most common are the spring jaw or C-clamp types. Many of the difficulties in arc starting and electrode sticking can be traced to poor grounding connections.

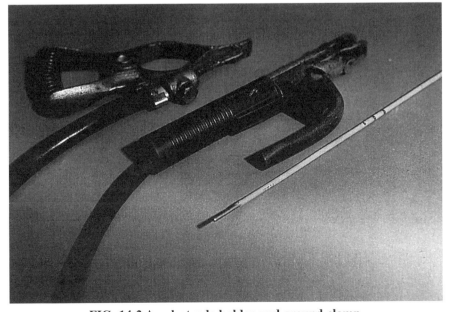

FIG. 14-2 An electrode holder and ground clamp.

Welding Current Selection

Depending on the type of welding power supply, the welder has the option of selecting from three types of current. Each current type has particular characteristics (when used with a suitable electrode) that produce optimum welding conditions.

1. Alternating current (AC): Alternating current is the most efficient arc, with higher arc temperatures per amperage setting. AC is the first choice for production welding on heavier plate thicknesses and is a cure for "arc blow" which results from a fixed polarity current that can magnetically deflect the arc away from the joint.

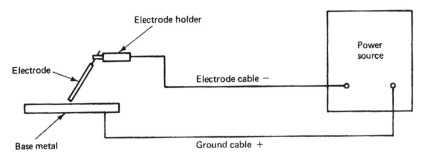

FIG. 14-3 A DC straight polarity welding circuit.

2. Direct current, straight polarity (DCSP; electrode negative) (Fig. 14-3): Direct-current, straight polarity is a smoother, less penetrating arc at any given amperage setting, producing less spatter than AC. It is the ideal choice for thinner material (sheet metal), wide gaps, and some out-of-position applications.

3. Direct current, reverse polarity (DCRP: electrode positive) (Fig. 14-4): This current produces deeper penetration at minimum amperages. It is the best choice for all out-of-position welding, tight fit-ups, welding on rusty or plated surfaces, and where minimum heat input to the work is desirable, such as when welding stainless steel and in repairing cast iron.

When the choice of current is limited to one type or another, the choice

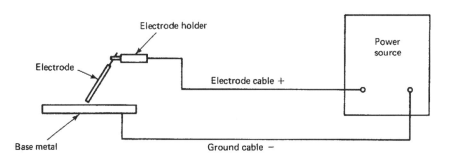

FIG. 14-4 A DC reverse polarity welding circuit.

of the electrode, together with the proper operator technique, will produce good, sound welds under a wide variety of circumstances. However, it will be the selection of a particular current type that will result in the best overall production efficiency.

SMAW Electrode Classification

The American Welding Society has established a classification system to enable the welder to determine electrode tensile strength, welding current, and handling characteristics of individual electrodes (Fig. 14-5). Although there is a classification for all types of electrodes and filler rods, this section will deal primarily with mild steel electrodes. Because the core wire of all mild steel electrodes is basically the same, variations in flux coatings determine their welding characteristics.

HOW THE CLASSIFICATION WORKS

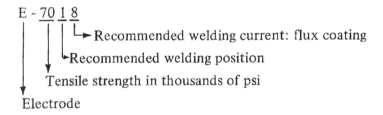

NOTE: If the classification number has five digits, the first three will will indicate tensile strength.

FIG. 14-5 Mild steel SMAW electrode classification code.

Welding Position Digits

1. All positions
2. Horizontal fillet and flat positions only
3. Flat only

TABLE 14-2 gives the last number of the code indicating the type of flux coating material and the recommended welding current type.

TABLE 14-2

Code Digit	Flux Coating	Current Type
0	Cellulose sodium	DCRP
1	Cellulose potassium	AC, DCRP
2	Titania sodium	
		AC, DCSP
3	Titania potassium	AC, DCSP
4	Iron powder, titania	AC, DCSP,
		DCRP
5	Low-hydrogen sodium	DCRP
6	Low-hydrogen potassium	AC, DCRP
7	Iron powder, iron oxide	AC, DCSP,
		DCRP
8	Iron powder, low hydrogen	AC, DCRP

Most electrode manufacturers now print the AWS classification number on the flux coating (Fig. 14-6). A trade name may also appear, but these are proprietary terms and while descriptive, it is the AWS code that fully and accurately describes the electrode in question.

FIG. 14-6 The AWS electrode classification number is printed on most brands today.

The Arc Welding Electrode

The shielded metal arc welding electrode is a metal rod with a flux coating used to carry and direct the welding current to the point at which the weld is made. Electrodes are available in a variety of types and sizes (see Appendix Section) and are consumable, meaning that during the welding operation the electrode is melted, mixes with the base metal, and becomes part of the weld. Many times the successful completion of a job is determined by the welder's ability to select and use the best electrode.

The size of the electrode is determined by the diameter of the core wire (Fig. 14-7). The most common packaging method used is the 50-lb box or can. The more expensive specialty electrodes are available in smaller sized containers. Electrode length may vary with the diameter, but generally it is 14 in. long.

FIG. 14-7 The electrode size is determined by its core diameter.

Purpose of the Flux Coating

♦ Cleans and deoxidizes the surface of the metal being welded.
♦ Provides easier arc starting and increased arc stability.
♦ Directs the force of the arc to permit maximum penetration and weld uniformity.
♦ Creates a shielding gas that protects the molten metal of the weld puddle from atmospheric contamination.
♦ Makes possible X-ray-quality welds.
♦ The resulting slag acts as an insulator and slows the cooling rate of the weld, resulting in greater weld bead ductility.
♦ Mixes with the molten metal and helps float impurities to the surface where they become part of the slag and are removed.
♦ Determines the basic operational characteristics of the electrode.

Mild-steel electrodes are classified in one of the following groups:

1. **Fast freeze:** Electrodes in this group (E-6010, E-6011) are designed to be used in general fabrication, maintenance, and pipe welding. The bead sets up very quickly, which makes these electrodes excellent for out-of-position welding. The welds produced have deep penetration, a light slag covering, and fair bead appearance.
2. **Fast fill:** Electrodes in the fast-fill group (E-6027, E-7024) are designed primarily for high-speed, flat-position welding of tight-fitting joints. The flux coating has a high iron powder content, which produces a weld with a smooth bead, excellent appearance, and shallow penetration. These are high deposition electrodes with easily removed slag coverings.
3. **Fill freeze:** Fill-freeze electrodes (E-6012, E-6013, E-7014) are excellent general-purpose rods. They can be used in all positions and produce a bead with medium penetration and good appearance. They are also recommended for the welding of sheet metal.
4. **Low hydrogen:** The most commonly used low-hydrogen electrode E-7018, is designed to produce a high-strength, X-ray quality weld. It is used extensively in the construction industry when high-quality plate or pipe welding is required. It can be used in all positions and produces a weld bead with good appearance and moderate penetration. This electrode is also used for welding low alloy, high strength steels. The 7018 can also be used when welding heavy sections of mild steel that are susceptible to underbead cracking as they cool.

Using these group characteristics as a reference, the selection of a specific electrode is based upon the type of metal to be welded, the welding position, type of current available, and the fit-up and geometry of the joints. In this last consideration, a deep penetrating electrode may be required. In other cases, welding speed may be of prime importance dictating the use of "fast fill" electrodes.

Proper electrode handling dictates the following procedures:

♦ Handle electrodes carefully to avoid damage to the flux coating.
♦ Store electrodes in a dry area.
♦ Open boxes should be kept in an electrode oven.
♦ Avoid contaminating the flux coating with water, oil, grease, etc..
♦ The low-hydrogen electrodes are best used hot (200-250°),
  especially for code work. They should never be left lying about
  where they can pick up moisture in their coating.

Not all electrodes require holding oven temperatures, but all should be kept dry and prevented from absorbing moisture. A simple plywood box fitted with a 100-watt light bult is adequate to maintain electrode condition and dryness. Table 14-3 provides recommended holding temperatures.

Current Setting and Arc Length

The determination and setting of welding power supply output current (amperage) is a key component of successful arc welding. Welding amperage is

Electrode Classification	Recommended Storage Unopened Boxes	Recommended Storage Open Boxes	Holding Oven	Reconditioning
E-XX10	Dry @ room temp	Dry @ room temp	Not recommended	Not done
E-XX11	Dry @ room temp	Dry @ room temp	Not recommended	Not done
E-XXI2	Dry @ room temp	Dry @ room temp	Not recommended	Not done
E-XX13	Dry @ room temp	Dry @ room temp	Not recommended	Not done
E-XX14	Dry @ room temp	150-200 °F	150-200 °F	250-300 °F
E-XX20	Dry @ room temp	150-200 °F	150-200 °F	1 Hour
E-XX24	Dry @ room temp	150-200 °F	150-200 °F	
E-XX27	Dry @ room temp	150-200 °F	150-200 °F	
E-60 or 7015	Dry @ room temp	250-450 °F	150-200 °F	500-600 °F
E-60 or 7016	Dry @ room temp	250-450 °F	150-200 °F	1 Hour
E-7018	Dry @room temp	250-450 °F	150-200 °F	
E-7028	Dry @ room temp	250-450 °F	150-200 °F	
E-80 & 9015	Dry @ room temp	250-450 °F	200-250 °F	600-700 °F
E-80 & 9016	Dry @ room temp	250-450 °F	200-250 °F	1 Hour
E-80 & 9018	Dry @ room temp	250-450 °F	200-250 °F	
E-90 12015	Dry @ room temp	250-450 °F	200-250 °F	650-750 °F
E-90-12016	Dry @ room temp	250-450 °F	200-250 °F	1 Hour
E-90-1 2018	Dry @ room temp	250-450 °F	200-250 °F	
E-XXX-15 or 16	Dry @ room temp	250-450 °F	150-200 °F	450 °F
Stainless	Dry @ room temp	250-450 °F	150-200° F	1 hour

TABLE 14-3 Holding oven temperatures.

directly proportional to the size of the electrode being used. In other words, larger electrodes will require a higher amperage setting than smaller ones. An easy rule of thumb to use in determining current setting is to adjust the power supply so that the amperage is approximately the same as the decimal equivalent of the electrode diameter.

Example: 1/8-inch electrode (0.125): 125 ampere initial setting
             5/32-inch electrode (0.156): 156 ampere initial setting

A given electrode diameter is designed to operate over a fairly wide range of amperage settings. For example, a 1/8 (0.125) inch diameter electrode will perform at amperages ranging from 85 to 145 amp. The exact and ideal final current for a specific instance is dependent upon several factors including:

- The type of electrode - lightly coated electrodes such as 6010 will require less amperage than the same diameter 7024 electrode with a heavier iron powder coating.
- The joint fit-up. Open joints are more easily bridged at lower amperages.
- The joint geometry - a "tee" joint in the center of a structure will require more amperage than an outside corner joint near an edge or cutout.
- The welding position - vertical and overhead welding and other situations in which gravity is a factor must be welded with lower amperage to reduce the liquidity of the weld puddle.
- The skill of the welder - more experienced welders will weld using the higher end of the amperage scale. The practiced welder is able to hold a closer arc length and is maniplitively more adroit.

Arc Length

The single most important factor for successful arc welding is the arc length. This lenth is defined as the size of the gap between the end of the electrode and the surface of the metal being welded (Fig. 14-8).

A general rule for most mild steel electrodes is that the arc length should be slightly less than the diameter of the electrode being used. There is no practical way for the welder to measure this length, so it must be determined by sound. The proper arc length will produce a sharp crackling sound similar to that of frying eggs. If the arc length is too short, the electrode will fuse itself to the weld puddle. If too long, the result will be poor bead formation and excessive spatter.

Select a piece of scrap steel and several electrodes. After the current setting has been made, run a weld bead and deliberately alter the arc length. Pay particular attention to the change in sound. After each bead, try to correlate the weld with the sound and take note of changes in weld appearance and excessive spatter. As the arc length increases, the sound changes from the sharp crackling noise to a hollow sounding hiss. As the arc length shortens, the pitch of the crackling increases. The welder's ability to maintain a proper arc length determines not only weld quality but appearance as well.

Starting to Weld

Use 1/4 or 3/8 thick hot rolled steel plate, about 6 inches square to allow several practice beads to be run before the plate becomes too hot. As practice continues, quench the plate in a tank of water so radical changes in amperage settings will not have to be made. Initial practice should be with 1/8 dia 6013 electrodes operating on AC current. The 6013 electrode has always been the easiest rod to learn with. When a professional looking bead can be run with 6013, practice may continue with other electrode types. Remember to reset the power supply current type and amperage to reflect the electrode being used.

Welding Technique

Place the electrode in the holder and position yourself so that you will be able to hold the electrode comfortably at approximately a 75-degree angle to the plate (Fig. 14-9). Heavier coated electrodes like the iron powder types (7024) may operate better at a smaller angle, and the lighter coated electrodes (6010) should be held almost vertical to the work. SMAW is always performed backhanded. A right-handed welder would weld from left to right and a left - handed welder welds from right to left. The easiest method of establishing an arc is by using a scratching rather than a poking motion with the electrode (Fig. 14-10), to minimize the problem of electrode sticking.

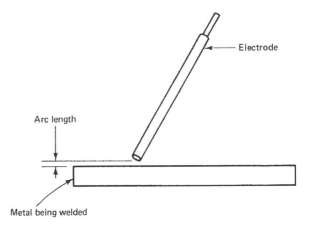

FIG. 14-8 Arc length is measured from the electrode tip as it is consumed to the surface of the work.

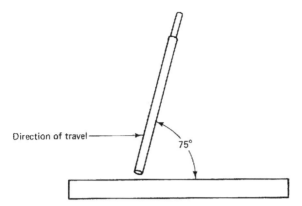

FIG. 14-9 The proper electrode angle to the work is approximately 75 degrees. this angle allows arc energy to push the slag away from the puddle.

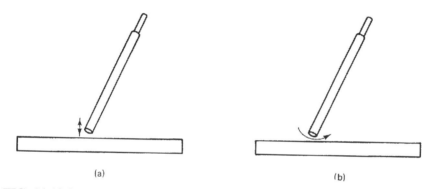

FIG. 14-10 A scratching motion will reduce electrode sticking when initiating the arc.

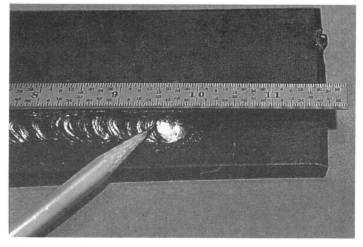

FIG. 14-11 Proper welding speed is shown by the center of the crater never being more than 1/4 in. ahead of the last bead nipple.

The next step is to run a weld bead. Strike an arc and move slowly across the plate, maintaining a proper electrode angle. While paying particular attention to arc length, listen for the sharp crackle. Remember, the electrode will become shorter during the welding operation and you must compensate for this by gradually moving it toward the plate surface to maintain a proper arc length.

Most beginning welders weld too fast. Welding should progress at a speed such that when the arc is broken the center of the crater should be about 1/4 inch ahead of the last ripple (Fig. 14-11). An indication that the proper amperage is being used is given by the weld puddle widening out to about twice the electrode diameter immediately upon the arc being struck.

Electrode Manipulation and Weave Patterns

Beginners will tend to weave the electrode without any distinct rhyme or reason, possibly believing that they are accurately imitating a more experienced welder. Such electrode manipulations have a purpose in out-of-position welding, and where some slight side-to-side or back and forth whipping motions are essentially compensations for some minor machine setting error. Ideally, the welding should progress from point A to point B with a straight line drag progression. But in practical shop and field welding, the same accurate amperage setting arrived at after many practice welds is not so easily set. As mentioned, the first indication of a proper amperage setting is when, immediately upon striking the arc, the puddle instantaneously widens out to about two rod diameters. If it does not, some side-to-side movement is permissible as the weld progresses. On the other hand, if the amperage has been set too high, the proper reaction is to "choke" the arc by holding a very short act length and whipping the electrode in short forward and back strokes. Care must be exercised to avoid welding over solidified slag. Movement back into the puddle must occur while the covering slag is still molten so that it can be washed out by the force of the arc.

When making fillet welds the same "from A to B" movement is most desirable in the flat or horizontal position. However, increasing the rod angle away from the vertical member and employing some side to side movement can control undercutting and improve bead appearance. An electrode should never be weaved any more than three times its diameter because excessive weaving will cause bead rollover and slag entrapment. If a larger fillet is required, use a larger diameter electrode. Low-hydrogen electrodes, which use molten flux to shield the weld puddle, should always be applied as stringer passes. If some weaving is necessary for out-of-position welding or as an amperage setting compensation, it should be done carefully.

Different weld bead types or weld passes are defined by their function in heavy plate fabrication and pipe welding and when material thicknesses are 1/4 inch and heavier.

♦ Root Pass - fuses the bottom part of the "Vee" joint to a backing strip or if an open root joint, penetrates right through, leaving a bead of weld reinforcement on the back side.
♦ Hot or Fill Pass - completes the filling of the root opening.
♦ Stringers - successive narrow beads that fill the joint up to or slightly below the surface of the work.
♦ Cap or Finish Pass - a wider pass usually weaved to complete the weld.

Some variations of the Cap Pass are shown in figure 14-12. Other examples are depicted in the Practical Welding Exercises that conclude this chapter.

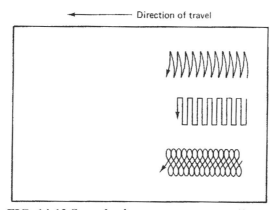

FIG. 14-12 Some basic cap pass weave patterns.

As stated, most initial practice should be with a 1/8-inch diameter, 6013 electrode. The 6013 electrode is an easy starting electrode producing moderate slag that is easily removed and is generally more forgiving in arc length and arc action. The following tips will facilitate practice with some other electrodes.

♦ 6010/6011 - Use DCRP with 6010 and AC with 6011. Requires 10-20 percent less amperage than 6013. Weld with the electrode held near vertical with a slight leading angle.
♦ 7024 - Operates on DCSP but AC gives the best performance. Requires amperage at the highest end of the range for a specific diameter. The leading angle while welding should be 30 to 40 degrees off the vertical.
♦ 7018 - Can be run on either DCRP or AC with DCRP being preferred in most situations. Initial amperage should be at about the middle of the recommended range. Electrode manipulations should be held to a minimum because shielding of the weld puddle is effected by slag coverage rather than any gaseous shield as with other electrodes. Low-hydrogen rods can be difficult to restart. Avoid banging of the rod end to remove the flux

which has solidified over the rod end. The proper method is to remove the rod from the holder and "wipe" the tip to expose the metal core. In using the low hydrogen electrodes there are some additional considerations. For best results they should be used slightly hot to the touch for code welding and at least warm for general work. Every effort should be made to prevent moisture pick up in the coatings (refer to Table 14-3). Welders who are experienced in using 6010 or 6011 will find that using a low-hydrogen electrode properly will require some practice. Out of position welding can be particularly challenging. Excessive "whipping" must be avoided. Many experienced Lo-Hy welders, when welding vertically, prefer to angle the elec trode down about five degrees while progressing upwards. Refer to figure 14-17 in which the conventional electrodes are shown pointing upwards at 5 to 10 degrees.

With any electrode, vertical-down welding is used only to create a smooth cap pass or to avoid burning through thin sheetmetal. In order to obtain good penetration and sound fusion, vertical welding should always proceed from bottom to top.

When welding stainless steel, electrodes must be selected according to the base metal characteristics. Electrodes designated XXX-15 are for DCRP, while the -16 electrode is for AC current. DCRP is preferred for its lower heat input to the work. Additional stainless data is to be found in the Appendix.

This chapter concludes with several practice exercises. Do not rush to practice the more difficult vertical and overhead welds until the flat and horizontal versions are mastered. Strive for each bead to be smooth, evenly rippled,

FIG. 14-13 Before practising with other electrodes and in positions other than flat, concentrate on the basic joints.

with no undercutting or excessive rollover (Fig. 14-13). Out of position welding will require closer arc lengths at reduced amperages so "sticking" will be a problem for the beginner. Arc length, rod angle, and welding speed must all be coordinated by the welder based upon a visual estimation of what is going on within the arc while welding. If, after chipping off the slag, the welder is surprised at the bead appearance, either good or bad, more practice is needed. **Practice material thickness should be 1/4 or 3/8 inch.**

Making a Fillet Weld (Fig. 14-14)

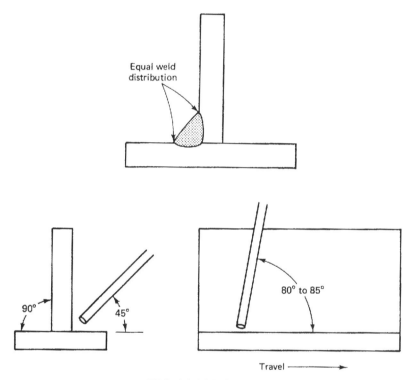

FIG. 14-14 A fillet weld.

Making a Butt Weld (Fig. 14-15)

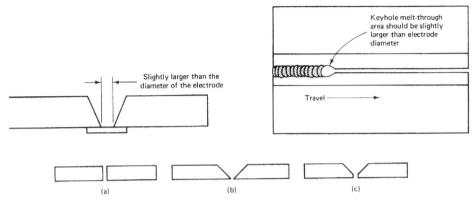

FIG. 14-15 A butt weld.

Welding in the Horizontal Position (Fig. 14-16)

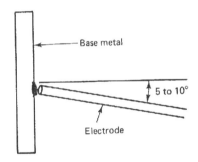

FIG. 14-16 Horizontal welding.

Welding in the Vertical Position (Fig. 14-17)

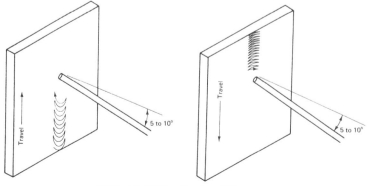

FIG. 14-17 Vertical welding.

Welding in the Overhead Position (Fig. 14-18)

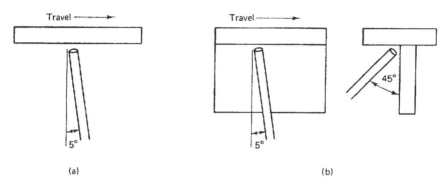

FIG. 14-18 Overhead welding.

Pipe Welding (for advanced students only) (Fig. 14-19)

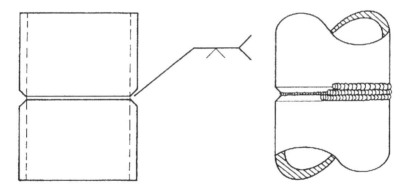

FIG. 14-19 Pipe welding (rolled or fixed positioning).

Common SMAW Weld Defects and Causes (see Table 14-4)

Welding Cast Iron
General Information

1. Types of cast iron:
   a. White: not recommended for welding
   b. Gray: most common
   c. Malleable: strongest of all types, seldom needs repairs
2. Types of electrodes:
   a. Machinable: nickel-core wire
   b. Nonmachinable: mild steel-core wire

Defect/Problem	Cause
Erratic arc	1. Loose cable connections 2. Poor ground 3. Current set too tow 4. Arc blow
Excessive spatter	1. Current set too high 2. Damp electrodes 3. Wrong polarity 4. Arc length too long
Poor bead appearance	1. Incorrect current adjustment 2. Faulty electrode 3. Electrode held at the wrong angle 4. Improper electrode manipulation
Poor fusion	1. Current set too low 2. Improper electrode size 3. Welding speed is too fast 4. Improper joint preparation 5. Improper electrode manipulation
Arc blow (erratic uncontrollable arc wander)	1. Magnetic fields in the weld joint[a]
Poor penetration	1. Current set too low 2. Welding speed too fast 3. Improper joint preparation
Porosity	1. Joint not properly cleaned 2. Insufficient puddling time 3. Improper electrode selection
Cracking	1. Weld bead not proportional to the base metal thickness 2. Improper electrode selection 3. Crater not filled 4. Joint not properly preheated 5. Current set too high

TABLE 14-4 SMAW problems and defects.

Preparation

2. Thoroughly clean the weld area to remove oil, grease, paint, and surface scale.
3. Grind a bevel on all broken edges (Fig. 14-20). Care should be taken not to bevel completely through the edge. Leaving part of the broken section intact will aid in realignment.

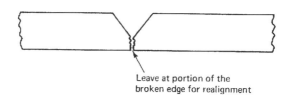

Leave at portion of the
broken edge for realignment

FIG. 14-20 A partial bevel will permit realignment of a broken casting.

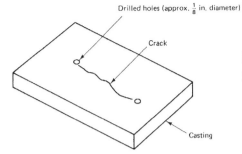

FIG. 14-21 Drilling a "stop" hole will prevent a crack from running.

4. If the casting is just cracked, drill a small diameter hole slightly beyond each end of the crack (Fig. 14-21). This hole will prevent the crack from traveling during the welding operation. V-groove the crack with a grinder.

Welding Technique

1. Preheat the entire casting to a temperature of 500 to 900 degrees Fahrenheit (°F). At these temperatures the casting will be smoking hot but not glowing.
2. Use DCRP if possible, with just enough amperage to operate the rod and fuse the joint.
3. Use medium to long arc lengths to avoid flux entrapment.
4. During welding, be careful to maintain the preheat to prevent any further cracks from developing. If possible, do all welding in the flat position. Weld on the opposite side to reinforce the joint.
5. If the repaired area is to be drilled or machined, a machinable-type nickel-core wire electrode is be used; alternatively, a type 308 or 310 stainless electrode can be used.
6. After repairs are completed, allow the casting to cool slowly. A light peening will reduce the chance of cracking due to bead shrinkage.
7. If preheating is impractical, weld the casting in 1-inch segments. Allow each segment to cool to the touch before proceeding to the next. Use the back-step welding technique (Fig. 14-22).
8. Retard the cooling rate by placing the repaired part in an oven, burying it in sand, or wrapping the part in a flameproof blanket.

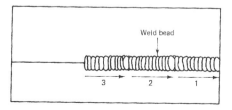

FIG. 14-22 Where preheating is impractical, use a back step welding sequence allowing each short weld to cool before continuing.

CHAPTER 15

GAS TUNGSTEN ARC WELDING

Process and Equipment

The gas tungsten arc welding (GTAW) process, or TIG as it is commonly known, utilizes an electric arc that is struck between the work and a nonconsumable tungsten electrode. The arc is contained within a shield of a chemically inert gas, usually argon (Fig. 15-1). Advantages of the process include rapid melting of the weld joint reducing overall distortion and adverse metallurgical changes in the heat-affected zone of the weldment. The weld quality is improved due to more effective shielding from atmospheric contaminants. For the welder, the TIG process affords improved visibility due to the elimination of smoke, spatter, and more welder control of variables such as penetration, bead size, and profile. **For practice, joints of similar size and thickness, as shown in Chapter 13 (OAW) can be used. However, cold-rolled rather than hot- rolled steel should be selected. Additionally, practice pieces can be stainless steel and aluminum.**

Power supplies designed for the TIG process have a high frequency circuit superimposed on the regular welding current. This enables "non-touch" starts to be made without contaminating the electrode.

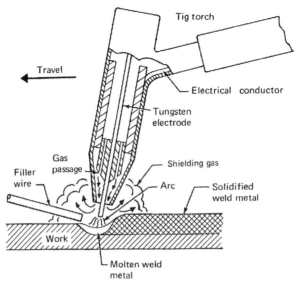

FIG. 15-1 The gas tungsten arc welding process. (Courtesy Hobart Institute of Welding Technology)

The process is most efficiently accomplished with power supplies and accessories designed expressly for tungsten-inert gas welding (Fig. 15-2). Such equipment includes:

- A transformer-type welder with coil windings significantly more massive than those of conventional arc welding machines. The machines are preferably AC/DC rectifier types with additional high frequency circuits built in.
- A torch and hose assembly with collet devices for holding the tungsten electrode, conducting shielding gas to the weld puddle, and providing a coolant for the torch head.
- Ceramic nozzles for directing the shielding gas flow.
- A supply of inert gas.
- A regulator flow meter that will indicate cylinder pressure in psi and shielding gas flow rate in cubic feet per hour (CFH).
- A separate reservoir and pump for the torch coolant. Air-cooled torches may be used for light duty production welding (100 amps or less). In the field, 150 amp air-cooled torches are regularly used.

Crucial to GTAW performance and productive capability is the welding machines duty cycle. As explained in Chapter 4, duty cycle is the number of minutes out of ten the welder can be operated at its full rated output without serious overheating. This is particularly true when TIG welding with AC current which, at 60 cycles, changes direction 120 times per second, thus "alternating" between DC straight polarity and DC reverse polarity. As the tungsten electrode used in GTAW is not consumed, there is no heat loss or dissipation. During the

FIG. 15-2 TIG welding is most efficiently done with power supplies designed from the process.

DC portion of the sine wave, this heat is returned back to the power supply to be absorbed by the heavier windings and insulation that characterize a power source designed for TIG. If a AC/DC power supply which is used for regular stick welding is subsequently adapted for TIG by the addition of a high frequency starting circuit and other controls, it must be derated as to it's duty cycle by at least 50 percent.

Another aspect of the newer models designed for TIG welding are the wave balance controls which have been developed. Figure 15-3 shows the basic AC sine wave and helps illustrate the principles behind GTAW current selection for various types of non-ferrous welding with AC current. AC current is a combination of DC straight and DC reverse current. The figure is a sine wave that has been rectified to produce DC current which through a series of rectifiers and other circuitry gating devices can be either DCSP or DCRP. Both diagrams are

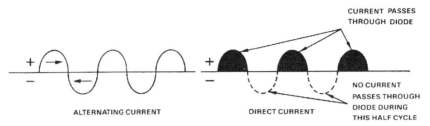

FIG. 15-3 The AC sine wave is rectified to produce DC current.

"balanced waves" in that each peak-to-peak valve is equal. Early TIG welders featured a balanced wave or a wave slightly unbalanced in favor of the DCRP portion. With the electrode positive and the work negative, the current flow from negative to positive would "electrically clean" the surface oxides that formed on the surface of aluminum and magnesium. Such an unbalanced wave was more efficient in breaking up and lifting the quickly forming oxides from the work surface. The down side was that the tungsten electrodes ran hotter and were more prone to excessive balling and spitting. The wave balance control shown in figure 15-4 enables the welder to customize the wave form to the exact needs of the job, especially when welding very thin materials, and when using small diameter electrodes.

FIG. 15-4 The wave balance control will adjust the DC component peak values of the AC wave.

Power Supply Setup

The arrangement of controls, which enables the welder to quickly set the machine for particular welding operations, will vary according to the brand name being used (Fig. 15-5). However, the basic complement would include:

♦ Off/ready switch, with no welding current flowing until a supplemental contactor is closed
♦ Mode switch for selecting either "TIG" or "STICK" processes

FIG. 15-5 Typical control panel arrangement of a TIG power supply.

♦ Shielding gas post-flow timing control
♦ High frequency three-position switch: continuous, off, and start
♦ High frequency intensity and phase-shift controls
♦ Current polarity switch
♦ Wave balance control
♦ Current range switch for determining amperage output within a specified range
♦ Current adjustment rheostat
♦ Panel/remote switch to determine the location of current adjustment
♦ Pulse controls (optional)
♦ Program and weld schedule controls (optional)

To set up the panel for TIG welding non-ferrous material set the mode switch to "TIG" and the high frequency switch to "continuous." This will maintain the arc as the current goes through the "zero" portion of the sine wave. Set the current selection to "AC." When welding steels, set the current selector to DCSP. Reverse polarity is never used in manual TIG although it has been used successfully in mechanized welding of very thin material. The high frequency control is set on "start" as the high frequency current is only needed to initiate the arc.

When TIG welding, if a remote foot pedal or torch-mounted control is used to control amperage output, the current adjustment rheostat is turned to its maximum setting. The panel/remote switch should be set to "remote." When the panel/remote switch is set on "panel," current output is governed by the setting of the panel rheostat. With a remote current control the panel rheostat can be used as a limiting device by setting less than the maximum for that range.

Other control variables set by the welder are the high frequency phase and intensity controls and post-flow timer. The high frequency circuit should be adjusted to produce a smooth AC arc. The post-flow timer should be adjusted so that the shielding gas flows long enough after the arc is broken to protect the tungsten electrode while it is still glowing. A 3/32-.dia tungsten elrctrode will require approximately a 10-second post-flow.

Regular Stick Welding

When using the TIG power supply for SMAW or "stick welding" the settings should be as follows:

♦ Mode switch on "STICK" or "SMAW." In this position an internal supplemental contactor is closed, making welding current immediately available at the electrode holder.
♦ High frequency switch to the "off" position.
♦ Current polarity switch set as required.
♦ Current range switch set as required.
♦ Panel/remote switch on "panel."
♦ Current rheostat to whatever setting the job requires.

Characteristics and Application of Current Types

In the GTAW process, three welding currents are available: (a,) alternating current, referred to as AC; (b,) direct-current reverse polarity, or DCRP; and (c,) direct-current straight polarity or DCSP. Each of these currents produces a distinct pattern of weld penetration depth and width (Fig. 15-6).

Three basic concepts of non-consumable electrode welding come into prominence at this point. First, in the GTAW process, no metal is transferred across the arc. Second, in DCSP welding, the work is the positive pole of the circuit, whereas in DCRP welding, the electrode is the positive pole. Third, in a GTAW arc, 75 percent of the heat generated is at the positive pole.

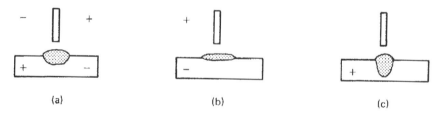

(a)　　　　　　　　(b)　　　　　　　　(c)

FIG. 15-6 Profiles of penetration for TIG welding currents.

DC straight polarity is used in welding of steel and copper alloys. It produces a bead of narrow width and deep penetration. Metals with an oxide coating such as aluminum and magnesium are welded with AC and continuous high frequency stabilization. Recall that AC is actually a combination of DCSP and DCRP and the reverse-polarity phase is an aid in removing those oxides. Wave balance adjustment will provide very fine control if available. DCSP current may cause some burn-through on very thin materials like copper and brass. In these instances AC with continuous high frequency may be used in conjunction with some wave balance control.

Tungsten Electrodes

Tungsten electrodes used in GTAW come in a variety of types and diameters (Fig. 15-7). As "nonconsumable" electrodes, they are virtually indestructible during the welding process. However, because of the high cost of these electrodes, poor welding technique and careless handling will prove to be prohibitively expensive.

The basis for selection by diameter is based on the welding amperage the tungsten would be expected to carry. Too much amperage for a given diameter will cause melting of the tungsten and contamination of the weld puddle. Too little current will result in excessive "arc wander" and poor puddle control.

Selection by Type

♦ Pure (color code green): For continuous production welding of aluminum and magnesium. It is easier to form and maintain the spherical end (ball) required in AC/HF welding.

♦ Thoriated (1 percent thoria, color code yellow, 2 percent color code red): For DC welding of steel alloys, brasses, and copper alloys. These types may also be used on aluminum and to reduce inventory costs in shops that weld a variety of metals. Thoriated electrodes have better arc starting characteristics and smoother operation at low amperages.

♦ Zirconium (color code brown): These electrodes are premium-type and are for use where the smoothest AC arc action is desirable while eliminating even the most minute amount of tungsten loss into the weld puddle.

FIG. 15-7 Due to their cost, tungsten electrodes require careful handling and use.

In addition to the basic tungstens discussed there are some specialized varieties that can be considered. "Ceriated" tungstens can be called a true "premium" electrode for both DC and AC welding. They have a higher current carrying capacity and will maintain their tip preparation for a longer period. Other tungstens are of the "lanthanated" variety. They have very low electrical resistance for easy starts and will carry twice the amperage of an equivalent sized pure tungsten. These are very expensive tungstens and their use may not always be justified.

The standard purchased length for use in manual torches is 7 inches, and it is desirable to use them as such. If breaking the tungsten into shorter lengths proves necessary, they must first be "scored" with a grinding wheel, then cracked by hand. If treated otherwise, they often "shale" and split into unusable pieces.

If the tungstens become contaminated during welding, they may be ground smooth on a fine grinding wheel. Grinding marks should be made parallel to their length (Fig. 15-8). Tungstens with severely contaminated ends should be broken off and redressed.

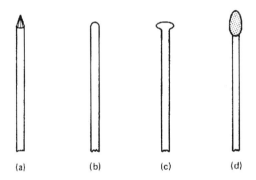

(a)   (b)   (c)   (d)

Contaminated and properly dressed tungsten electrode tips: (a) 2% thoriated tungsten ground to a taper for DCSP welding; (b) pure tungsten with a ball end for AC welding; (c) tungsten with an excessive ball end; (d) tungsten badly contaminated having repeated contact with filler rod or weld puddle.

FIG. 15-8 Contaminate tungsten's can be easily redressed.

Inert Shielding Gases

The function of the shielding gas is to exclude the contaminating atmosphere from the molten weld puddle, assuming that cleaning of the material prior to welding has been completed. The gases used are described as "inert" in that they will not interact with the molten weld metal to form chemical compounds harmful to the weld. In manual gas tungsten arc welding argon is the most commonly used; however, helium and argon/helium mixtures also have desirable characteristics.

♦ Argon will provide the smoothest arc action, aiding in bead formation and penetration control. Although equal in cost to helium, argon is more economical to use. Being heavier than air, argon allows the use of lower flow rates. Easier arc ignition and wider variations in arc length are also possible with argon.

♦ Helium will effectively raise arc temperature at any given arc length and amperage setting. Helium allows faster welding speeds and deeper penetration, but with significant loss of arc stability. Helium finds its greatest use in welding nonferrous material 1/4 inch and thicker.

♦ Argon/helium mixtures, generally 75 percent-25 percent helium, will maintain appreciable arc stability while still provide higher arc temperatures. Argon/helium mixtures can limit the need for larger-diameter tungstens and torches of higher capacity.

♦ Argon with up to 5 percent hydrogen is used for manual welding of the austenitic 300 series, stainless steels. This mix, because of the hydrogen, cannot be used when welding carbon steels, aluminum, titanium, and other reactive metals. For most applications however, straight argon can be used.

Inert gas flow rates are a function of the work thickness, tungsten diameter, and nozzle size. Higher flow rates are required by longer-than-normal tungsten extensions (Fig. 15-9a and Fig. 15-9b). The figure also depicts the typical torch component assembly.

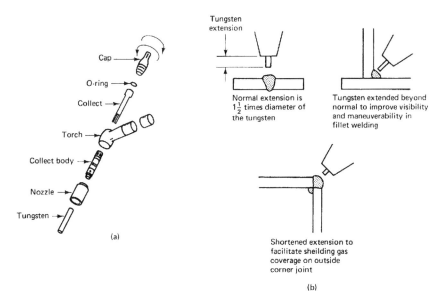

FIG. 15-9 "a" Torch assembly detail. "b" Longer tungsten extension will require an increase in shielding gas flow (CFH).

TIG torches come in a variety of sizes and current carrying capacities. Water-cooled torches are the most common but air-cooled torches are popular for portable field welding. Many torches can be fitted with micro switches for starting the arc and for current adjustment. Inert gas nozzles or "cups" are made of a high temperature resistant ceramic material and come in a variety of shapes and sizes. A "gas lens" has been developed which projects a longer pattern of gas coverage. This is useful allowing the tungsten to project up to 3/4 inch for better visibility in tight spaces and in outdoor welding where air currents could affect the gas coverage.

A table of tungsten diameters, nozzle or cup sizes, and shielding gas flow rates can be found in the appendix.

Welding Aluminum and Magnesium

- ♦ **Cleaning:** Aluminum has an oxide coating that must be removed prior to welding. This oxide does not melt during the welding process and will contaminate the weld. Removal is accomplished by immersion in a chemical etchant or by mechanical means such as wire brushing. Films of grease or wax can be removed with a non-petroleum-based solvent such as acetone.
- ♦ **Joint preparation:** Standard joint preparation practices are applicable to aluminum. However, certain aspects should be noted. Joints should be accurately fitted to minimize shrinkage of closely dimensioned parts and the joint design should allow for the maximum use of filler rod (Fig. 15-10). Unlike the steel alloys, fused welds with no filler will leave weak and crack-sensitive joints.
- ♦ **Current type:** High frequency stabilized alternating current (AC/HF). Wave balance adjusted as needed.
- ♦ **Current range:** When welding material 1/16 to 3/16 inch thick, the "medium" range position is ideal. The "low" position will provide more sensitive control on the thinner gauges, while the "high" range position is less fatiguing when welding at maximum amperage on thicker materials. The basic current control device is a foot pedal rehostat.
- ♦ **Voltage:** Manual welding is done with constant-amperage-type power supplies, so voltage variations are a direct result of arc length. Arc lengths of from 1/16 to 3/32 inch, steadily maintained, provide suitable voltage readings for good bead formation. As shown in Figure 15-11, the arc pattern is distinctly conical in shape. Arc-length variations not only influence arc voltage but actually increase or decrease the joint area being heated regardless of any amperage setting. Well trained and competent welders in the GTAW process can limit arc length variations to only 0.010 to 0.015 inch.

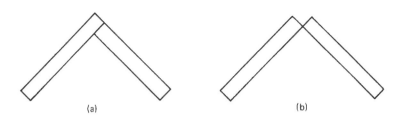

Sample joint design: (a) not recommended, excessive filler rod dilution; (b) preferred, maximum addition of filler rod provides highest strength and ductility.

FIG. 15-10 Joint designs for TIG should allow for the maximum addition of rod filler metal.

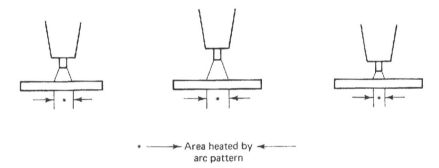

* ⟶ Area heated by ⟵
arc pattern

FIG. 15-11 Changes in TIG arc length will affect base metal heating.

♦ Shielding gas flow (argon): Most welded sheet metal fabrication up to 1/8 inch thick can be accomplished with flow rates of 15 to 20 CFH. Greater thicknesses require up to 30 CFH with suitably sized torches, tungstens, and nozzles.

For material up to 1/8 inch thick, a 3/32 inch diameter tungsten electrode can be used. If practice material is 1/16 in. thick, use either a 1/16 or 0.045 diameter electrode, either pure or thoriated tungsten. If the latter is used, the tip must be "balled (see Fig. 15-8). This result is accomplished by setting the power supply current selector on DCRP and momentarily causing a rapid melting of the tungsten. Larger weldments up to 1/4 inch thick will require a 1/8 inch diameter electrode and some mild pre-heating.

Filler Rod Selection (Aluminum)

The filler rod selected for use with a particular grade of aluminum must be metallurgically compatible with the parent metal being welded. Typical examples would include:

Parent Metal	Filler Rod
3003	1100
5052	5154
6061-T6	4043

(See the appendix for more detailed data)

The proper welding technique includes:

+ Weld in the forehand manner whenever possible.
+ Tungsten "stickout" from the nozzle should be from 1 to 1-1/2 times the electrode diameter.
+ Torch incline, in the direction of travel, should never be more than 15 degrees off perpendicular (Fig. 15-12).
+ Follow this basic sequence:
  - Create a molten puddle.
  - Add rod to the leading edge of the puddle.
  - Advance the puddle along the joint approximately one-half of its own diameter. Avoid weaving or any type of side-to-side torch movement.
  - Add rod again to the leading edge. Critical to the process is the ability to feed the filler rod steadily in to the puddle with the opposite hand. For this reason, a glove is not usually worn on the rod hand.
+ Regulate the amperage so that a welding speed of 12 to 18 inches a minute can be maintained without excessive buildup or under cutting.

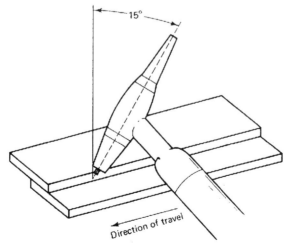

FIG. 15-12 To insure good shield gas coverage, torch angles should not exceed 15 degrees off perpendicular.

Additional safety considerations when welding aluminum include:

♦ There is no color change as aluminum reaches its melting point, so there is no way to judge its actual temperature. Assume that it is hot.
♦ The reflective nature of aluminum can cause severe arc burn to exposed skin.
♦ A darker lens shade may be required for continuous production welding.

The same basic machine settings are used when welding magnesium. Surface oxides must be removed and additional cleaning of the filler rod with stainless steel wool is also recommended. Joint preparation is similar to aluminum but the use of run-in and run-out tabs (Fig. 15-13) will overcome the tendency for magnesium to crack. Straight argon is used for shielding.

In order to control bead width more closely and to minimize arc pitting alongside the weld, a slight taper is ground on the tungsten before "balling." Thus, a thoriated-type tungsten should be used. Both wrought and cast magnesium can be effectively welded; however, care must be taken to select those alloy types that respond well to fusion welding and to use the correct filler rod. The magnesium alloy most commonly fabricated in sheet metal is No. A231B. Filler rods No. AZ61A or AZ92A may be used, with the latter leaving a more ductile deposit. AZ92A should also be used in repairing castings.

The welding technique is somewhat different from that employed on aluminum. Rather than the distinct dipping of the rod into the puddle as it is advanced along the joint, the rod should be laid almost flat along the joint while the advancing puddle melts the end of the rod. To avoid stress corrosion cracking alongside the welds, stress relieving is recommended. Arc lengths should be 1/16 inch or less. Magnesium castings will require pre-heat as well as some post-heat treatment.

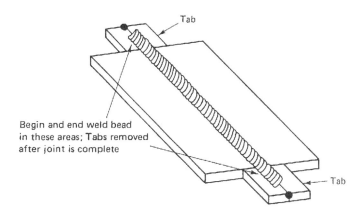

Tab

Begin and end weld bead
in these areas; Tabs removed
after joint is complete

Tab

FIG. 15-13 Run-in and run-out tabs will help eliminate the tendency for bead cracks in magnesium.

Magnesium welding has some additional safety concerns. Solid magnesium will not burn or flare at welding temperatures. However, finely divided particles such as shavings, chips, and grinding dust that may accumulate in and around welding fixtures are a definite and serious fire hazard. Good ventilation in any welding operation is always desirable, but is especially important with magnesium if the thorium alloys are being welded.

Welding Low-Carbon and Low-Alloy Steel

Hot rolled steels should be scraped and wire-brushed to remove any loose mill scale. Cold-rolled steels, which are preferred for GTAW fabrication, should be degreased with aviation turco or a similar non-petroleum-based solvent (acetone). Closely fitted joints are essential. Thinner-gauge sections can warp severely. Use backup and chill bars wherever possible. Butt joints over 3/32 inch thick should be beveled if full penetration is needed (Fig. 15-14).

Use direct-current straight polarity as a machine setting. The high frequency switch should be set on "start." At that setting, the high frequency is timed for about 5 seconds to initiate the arc, then turned off automatically. A "touch start" without high frequency is possible by a careful scratching of the tungsten. Short arc lengths and slow travel at minimum amperages will produce the best weld profile and appearance.

Low-carbon and low-alloy steels are generally of the "killed" or "rimmed" variety. Satisfactory welds on the killed type can be obtained with RG-45 gas welding wire. The rimmed type is not completely deoxidized during the refining process and is subject to excessive porosity. To avoid porosity and for superior ductility in both types of low-carbon steel, use E-70S-2 triple, deoxidized wire. For the higher-yield-strength alloys, use E-70S-4 or a rod as similar as possible to the parent material. Fused joints without rod are possible if the killed variety is used in fabrication and joints are carefully fitted and overlapped.

Straight argon gas is recommended for most applications, with flow rates of 8 to 10 CFH on joints up to 1/8 inch thickness. Usually, 2 percent thoriated tungstens are used, ground to a point as previously discussed. Amperages used will be 15 to 20 percent less than similar joints in aluminum. Avoid using tungsten of too large a diameter to eliminate arc wandering. Additional welding data can be found in the appendix.

Basically the welding technique is similar to that used for aluminum; form a puddle, advance the puddle by about one-half of its diameter, add rod, and advance again. Bead size should be kept to a minimum for two reasons. First, the low-carbon steels have yield strengths one-third higher than the strongest aluminum, and oversize beads are of no value and are wasteful. Second, exces-

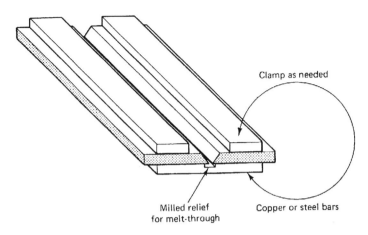

Clamp as needed

Milled relief
for melt-through

Copper or steel bars

FIG. 15-14 When welding sheetmetal thicknesses, chill and back-up bars will control warping

sive distortion will result as well as the tendency to undercut. Maintain very short arc lengths and add just enough rod to fill the puddle.

Welding Stainless Steel

TIG welding is the ideal process for welding the stainless steels. There is no spatter and minimum part discoloration. Beads are bright and smooth which is essential for efficient weld dressing. This is an important aspect as many stainless weldments are used in the medical, chemical, and food processing industries and all weld beads are ground smooth and flush.

Stainless steel is ready for welding as supplied by the distributor. The exception would be those sheets and shapes with brushed or polished finishes which have an adhesive paper covering. Any residual adhesive may be removed with acetone. Clean, closely fitted joints are essential. Use chill bars, clamped as close as practical to the joint seam, not only to limit distortion, but to keep heat discoloration to a minimum. The use of backup bars behind the joint seam will reduce the chance of burn-through. If the joint is to be fully penetrated, the melt-through must be shielded with inert gas (purging) (Fig. 15-15).

Use direct-current straight polarity. The high frequency switch is set on "start." It may be desirable to adjust the post-flow timer to provide a slightly longer flow duration after the arc is stopped. Stainless steel exhibits a relatively low thermal conductivity and high electrical resistance. Overheating due to excessive amperage can be avoided by using the panel rheostat as a limiting device. Most welded fabrication of stainless steel is done with austenitic chrome-nickel types such as 304, 308, and 316. Filler rods carry the same numerical designations and can be used without any pre- or post-heat treatment, and will leave deposits most similar to the parent material. Other types of stainless will require specific rod types and post-heat treatment. See the appendix for additional rod data.

Straight argon remains the usual choice for stainless steel regardless of plate thickness. When welding light-gauge material using small-diameter tungstens and nozzles, the argon flow should be lowered to 10 CFH to avoid turbulence over the weld area. Usually, the 2 percent thoriated tungstens are used, ground to a point as previously discussed. In certain more critical applications, zirconium-type electrodes may be used.

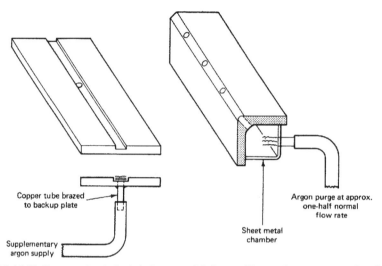

Copper tube brazed to backup plate

Supplementary argon supply

Argon purge at approx. one-half normal flow rate

Sheet metal chamber

FIG. 15-15 Fully penetrated stainless steel joints will require an argon back-up shield.

Welding technique for stainless steel is similar to that for carbon steel, but with some small modifications. To avoid sticking the tungsten and making inadvertent arc marks on highly polished surfaces, initiate the arc by resting the nozzle on the work at an angle that sets an adequate arc length. Start the welding sequence and as the arc is ignited, raise the nozzle and bring the torch to a more vertical position (Fig. 15-16). As mentioned, stainless has a high electrical resistance. This causes the weld joint area to become very hot, so much so that the crater and the last 1/4 to 1/2 inch of bead will be glowing red. To prevent contamination, hold the torch in position over the crater after the arc is broken and allow the post-flow of gas to protect the weld until it cools. In general, the use of multiple stringer beads is preferable to wide weaves to avoid excessive heat buildup in the weldment. The end of the filler rod should be kept within the inert gas flow as the welding progresses.

Welding Special Metals and Joining Dissimilar Alloys

The GTAW process excels in the fabrication of copper and copper alloys, brass, and bronzes. In addition, many of these can be welded to many carbon-steel alloys and to each other. Cast iron is effectively welded with ade-

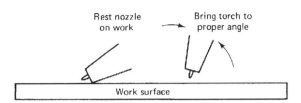

Rest nozzle on work

Bring torch to proper angle

Work surface

FIG. 15-16 By resting the gas cup on the work, there is less chance of making an inadvertent arc strike on the work.

quate pre-heat on large sections. Direct current straight polarity is used on all of the metals listed above. Materials 0.040 and thinner are best welded with AC/HF.

In general, for maximum ductility, use a rod that is similar to the base metal. Often, adequate results can be obtained with a copper-silicon rod known as Everdur. The drawback would be a lack of color matching in certain applications. Everdur will give a good color match with brass and has a high tensile strength. Also, Everdur is used in joining copper alloys to carbon and stainless steel as well as for cast-iron repairs. Argon is again the preferred shielding gas. Flow rates are generally between 15 and 20 CFH with material up to 1/8 inch in thickness. The 2 percent thoriated tungstens are best. The ground point will provide the greatest arc stability. In general, maintain the same manipulative sequences as in welding steels. Tight fit-ups and the use of backing plates and chill bars will keep distortion to a minimum.

Many of the metals and alloys discussed here will produce fumes that, if not toxic, are definitely irritating to the lungs and eyes. Adequate ventilation must be provided, especially when fusion-welding certain copper alloys and lead materials.

As mentioned earlier, TIG practice can be used on joints like those specified for oxyacetylene welding (Chapter 13). Remember to use cold rolled steel for DC practice and clean aluminum for AC welding.

Pipe Welding (Aluminum)

Machine settings, inert-gas flow rates, tungsten type, and preparation are identical as if welding sheet and plate. Welded piping is most often of the heat-treatable variety, such as 6061-T6. Pulsed arc control is desirable, especially on thin wall piping.

Pipe wall thicknesses of less than 1/8 inch may be set up as square butt joints (Fig. 15-17a). Thicker-wall joints should be chamfered 37 degrees on each side, with a nose of 1/16 in. tightly butted together (Fig. 15-17b). Pipe is either sawed or machined to size and fit. Lubricants used in such operations must be removed with the appropriate solvents. A welding filler rod leaving a heat-treatable deposit with good ductility such as No. 4043 is recommended.

Welding Technique

♦ Tungsten stickout beyond the nozzle end should be far enough to establish an acceptable arc length at the bottom of the groove. This will allow "cupping" of the torch along the groove walls while welding the root pass (Fig. 15-18).

♦ Root pass welding: Form a puddle across the joint. Dip the rod and hold the torch over the puddle until penetration takes place. This is determined by "reading the puddle." The puddle will flatten out and become wedge-shaped straight across the front and with rounded corners at the rear. Advance the puddle by one-half of its diameter and repeat the sequence. If the pipe is in the 2G position (vertical-fixed), the nozzle should be maintained slightly above the centerline of the joint.

♦ Filler passes: Weaved beads may be used on aluminum pipe without the problem of undercutting along the edges of the weld. However, in the case of overhead and vertical welding, the puddle must be more closely controlled to prevent crowning and sagging, which will be accompanied by excessive undercutting. Power supplies that can provide pulsed current are useful here. Also, as shown in Figure 15-19, the proper number and size of each fill pass must be planned so as to fill the joint flush or slightly below the surface of the outer wall of the pipe.

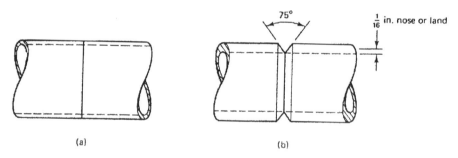

(a)   (b)

FIG. 15-17 a. Square butt joint for thin wall pipe. b. Thicker wall pipe is chamfered 37 degrees, leaving a 1/16 in. nose on land to complete the root.

♦ Finish pass: Cap welds should be about 1/8 inch wider than the top of the vee and evenly spaced on either side, about 1/16 inch higher than the surface. As the torch is weaved across the joint, rod is added when the tungsten is over the outer edges of the bead (Fig. 15-20).

Pipe Welding (Carbon and Stainless Steel)

Steel pipe TIG welding requires some special considerations, especially stainless pipe. The following techniques will ensure good quality joints that are far superior to similar pipe joints welded with SMAW.

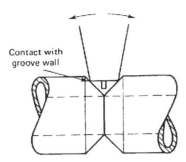

FIG. 15-18 "Cupping" is a method of establishing a correct arc length while going around a pipe joint.

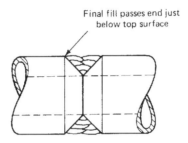

FIG. 15-19 Fill pass sequence for filling a pipe joint.

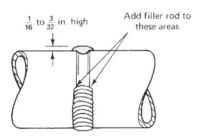

FIG. 15-20 The fine cap pass should be at least 1/8 in. wider than the top of the groove.

Bevel edges to 37-1/2 degrees. On carbon-steel pipe, remove varnish and rust by filing or wire brushing. Use a degreasing solvent if necessary. Set up the joint with a root opening equal to the filler rod diameter. Align and tack the joint in four places using the filler rod as a spacer. Weld deposits should be of the same mechanical properties as that of the pipe.

The "cupping" method described in the aluminum section may be used on the root pass. In root pass welding, hold the filler rod flat along the joint

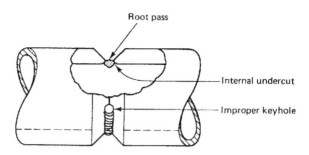

FIG. 15-21 When welding steel pipe, maintain an elongated puddle without "key-holing."

opening. This will prevent the deposit of unmelted portions of filler rod pro-truding through the penetration inside the pipe. Maintain a minimum bead width using just enough amperage to melt the rod and the edge of the joint. Avoid keyholing to prevent reverse undercut inside the pipe (Fig. 15-21). Run the torch over the end of the rod, melting it into the puddle. To obtain more penetration, carefully feed the rod into the puddle. Filler passes are welded by dipping the rod and advancing the puddle by about one-half of its diameter. Use stringers rather than weaves. Cap passes should be 1/16 to 3/32 inch higher than the pipe surface and at least 1/8 inch wider than the groove opening. This final weld should also be of multiple stringers.

If a consumable insert ring is used, adjust the current so that the ring will be completely melted and fused to the joint edges. This will ensure that the ring is following the inside contour of the pipe without bulging. Do not penetrate or add rod during this operation. Many field-welded piping operations may utilize standard DC welding machines not designed specifically for the GTAW process. Without remote contactors and high frequency it is necessary to initiate the arc by scratching the tungsten on the work. Care must be taken not to contaminate the crater. Using a backhand motion, strike the arc on the walls of the groove and then move it down into the joint (Fig. 15-22).

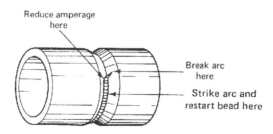

FIG. 15-22 If a "scratch start" is required, initiate arc on the beveled side walls not in the crater.

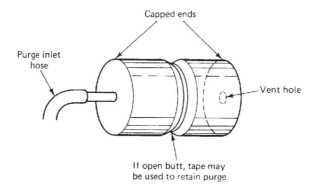

FIG. 15-23 Stainless pipe joints, with or without a consumable insert ring, require an internal gas shield or "purge."

When welding stainless steel pipe, with or without the use of a consumable insert ring, the melt-through must be protected from the atmosphere. This is accomplished by purging the inside of the pipe with argon. Flow rates of 5 to 7 CFH are adequate (Fig. 15-23). A heat-resistant tape may be used to seal off the balance of the root opening not being welded. After the root pass is complete, welding may proceed as outlined previously for carbon pipe. Although weaving may be acceptable on carbon pipe welded in the rolled position, it should never be attempted on stainless pipe. Excessive undercutting and overheating will result.

Common GTAW Weld Defects and Causes

Defect/Problem	*Cause*
Poor arc ignition and maintenance	1. Poor ground connection 2. Defect in high-frequency circuit 3. Tungsten diameter too large
Arcwander	1. Dirty tungsten 2. Deformed tungsten tip contour 3. Amperage too low for tungsten diameter
Tungsten contamination	1. Touching work with tungsten 2. Filler rod pickup on the tungsten 3. Inadequate post-flow
Dirty welds	1. Incomplete oxide removal from work 2. Oil or grease on work 3. Dirty rod 4. Inadequate shielding gas flow 5. Loose gas cup or torch cap 6. Contaminated shielding gas supply
Torch overheating	1. Excessive amperage beyond torch capacity 2. Constricted cooling-fluid flow 3. High-amperage welding in restricted joint areas
Porosity	1. Dirty base metal 2. Dirty rod 3. Inadequate shielding 4. Excessive welding speed 5. Excessive welding amperage
Lack of complete fusion	1. Amperage too low 2. Welding too fast 3. Improper joint preparation
Poor bead formation	1. Amperage too low or excessive for material thickness 2. Welding too fast 3. Improper choice of filler rod diameter 4. Improper torch angle to work
Undercutting	1. Improper torch angle to work 2. Improper size or application of filler rod 3. Weldingtoo fast 4. Excessive amperage
Cracking	1. Improper choice of filler rod type 2. Excessive dilution of filler rod with base metal 3. Too rapid chilling of weld joint 4. Highly restrained joints 5. Weld bead beginning on a tack 6. Unfilled craters

TABLE 15-1 GTAW defects and process problems.

CHAPTER 16

GAS METAL ARC WELDING

Process Description

Gas metal arc welding (GMAW), or MIG as it is commonly known, is a gas shielded arc welding process utilizing a continuous weld wire feed system. The process uses direct current, reverse polarity regardless of the type of metal being welded. As in the GTAW process, the weld takes place within a shield of inert gas (Fig. 16-1). MIG welding is commonly done in four basic forms: short arc, spray arc, pulsed arc, and globular transfer. These basic forms are known by a variety of process trade names by various manufacturers and essentially refer to the method of metal transfer across the arc. **Practice welding can be done on joints and material thicknesses as shown in Chapter 14 (SMAW).**

Power supplies

The power supplies used are classified as constant-voltage types as opposed to the constant-amperage types used in conventional arc welding. The voltage is set on the machine rather than the amperage. The power supply then has the potential for supplying whatever amperage the voltage across the arc calls for up to its rated output. These welding machines are also known as SVI power supplies because it is the combination of slope, voltage, and inductance that,

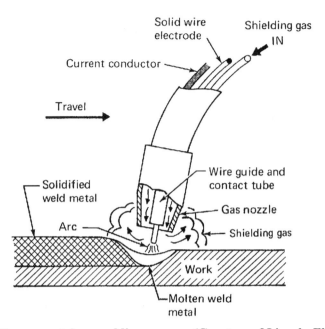

FIG. 16-1 The gas metal arc welding process. (Courtesy of Lincoln Electric Co.)

when properly set, provides for optimum welding performance on a wide variety of metals. Slope refers to the slant of the volt-ampere curve, indicating voltage change per 100 A. Whereas some machines have a fixed slope, more versatile types have variable slope settings. Those settings are either flat or steep and will be explained subsequently. Inductance in MIG welding can best be compared with fine tuning a TV set. The inductance controls the rate of the rise in the short-circuiting current without affecting the final welding current. Increasing the inductance slows down the current rise and reduces the tendency to spatter. Conversely, decreasing induction will allow a larger number of short circuits per second and reduce the fluidity of the weld puddle. The inverter power supplies have proven very efficient and versatile in controlling and fine tuning the GMAW process.

Wire Feeders and MIG Torches

Welding wire is supplied on spools of varying capacities. The wire feeder accepts the spool and conducts the wire through the connecting lines to the torch. A variable-speed motor, sometimes working in unison with a similar drive unit in the torch itself, feeds the weld wire into the puddle at a constant, predetermined rate. This wire feed rate, at a preset voltage, determines the welding amperage. Other systems made by various manufacturers are of the "pull" type, using a torch-mounted spool of wire, or of the "push" type, where the wire is merely pushed through a conduit to the torch or gun.

Additional Equipment

Additional equipment required includes a water cooling system for continuous welding in excess of 250 A, and the appropriate inert-gas cylinders, hoses, and regulating equipment. Many manufacturers offer "MIG sets" with matched components (power supply, wire feeder, and torch) designed to perform within the range of welding parameters dictated by specific production needs. There are MIG sets designed for light fabrication and body shop type work. Others are for intermediate duty and there are heavy duty industrial units for continuous production work.

Weld Metal Transfer Variations

Characteristics of the short arc method are:

♦ Lower voltages: 16 to 21 v
♦ Less heat input to the work
♦ Stiffer, less-fluid, puddles
♦ Marked tendency toward spatter

Characteristics of the spray arc method are:

♦ Higher voltages: 20 to 30 V
♦ Higher heat input
♦ More-fluid puddles
♦ Faster welding speed with virtually no spatter

Characteristics of the Globular Transfer Method are:

♦ Occurs with CO_2 shielding only
♦ Occurs with heavier wire diameters of 0.045 to 1/16 inch
♦ Produces deeper penetration with minimum spatter
♦ Requires higher amperage-to-voltage ratios

Characteristics of pulsed-arc transfer:

♦ Similar to spray arc, but with lower average current levels
♦ Frequency is about 60 pulses per second and it facilitates globular formation of molten wire before higher current pulse transfers the wire filler in a spray-arc mode
♦ Allows all-position welding and the welding of thinner materials with spray-arc characteristics.

The decision to use either method on a particular job is based on:

♦ The size and thickness of the work
♦ The position in which it must be welded
♦ Joint preparation and setup
♦ Type of shielding gas available
♦ Type and size of the welding wire available

Setting the Slope Characteristic (Variable-Type Power Supplies)

As mentioned "slope" refers to the volt-amp curve depicting voltage change per 100 amperes. A relatively steep slope of 3 or 4 volts per 100 amperes is ideal for short-arc welding. Flatter slopes representing 1 to 2-1/2 volts per 100 amperes are more applicable to spray transfer. Flux-cored welding will use a flatter slope for wire diameters over 1/16 inch and a steeper slope for thinner wires.

Setting Voltage and Amperage

♦ Depress the trigger on the welding gun. Set the voltage by turning the dial until the meter registers the desired value. Note that this "no arc" reading will be 2-1/2 to 5 V higher than actual welding voltage when the arc is struck
♦ Higher voltage settings will increase the heat intensity of the arc, thus allowing the use of thicker filler wire at higher feed rates. Welding amperages then will be higher. Lower voltages will, of course, have the opposite effect.
♦ A self- correcting arc length is a characteristic of the GMAW process. In conventional "stick welding," arc length controls the voltage across the arc. In MIG welding, it is the opposite. The preset voltage controls the arc length. Voltage settings that pro-duce arc lengths of from 3/32 to 1/8 inch are considered accept-able from the standpoint of puddle size and shape, spatter conditions and effective wire stickout dimensions. This acceptable arc length will be maintained automatically by the power supply throughout a narrow range of variations in the wire feed rate. Changes in wire feed that vary amperage within a 10 amp range can be made without requiring a corresponding change in voltage settings. Amperage changes beyond these 10 or so amperes, without

appropriate changes in voltage settings, will cause either:

    a. Shortening of the arc length to the point where spatter conditions become intolerable

    b. An increase in arc length to the point where a "burnback" occurs, fusing the wire to the contact tip

Setting Induction

- ◆ The induction dial is graduated from "minimum" to "maximum" in increments of 10. When the short arc mode is in use, effective set tings occur between "minimum" and "5". The higher reading (up to 5) will reduce excessive spatter to a marked degree. Lowering toward the minimum setting will reduce puddle fluidity, which aids in reducing burn-through on thin material and in bridging open root joints, and can be useful where minimum or at least controlled penetration is desirable.

- ◆ While welding in the spray arc mode, the most effective settings occur between "5" and "maximum." Slight increases above "5" will control spatter to the point where it is nonexistent. At higher wire feed rates, further increases will reduce "blasting," where the first 1/2 inch or so of wire embeds itself, unfused, into the weld deposit.

Flux-Cored Arc Welding

The flux-cored arc welding (FCAW) method is essentially a variation of the gas metal arc welding (GMAW) process. The power supply, the wire feeder, and the gun or torch, although similar in function to those used in GMAW, are usually designed specifically for flux-cored welding. The power supply, although of the typical constant-voltage type, has a fixed flat or medium slope. Provision for adjusting the wire feed speed is built into the wire feeder. The welding wire consists of a tubular sheath of mild steel containing a core of granular flux. In addition, a shielding gas may be used to enhance both the deposition rate and the quality of the weld bead. (Fig. 16-2).

FCAW finds its greatest use in welding a variety of steel alloys 1/4 inch and thicker. It is a high-deposition production process which calls for primarily flat-position welding of structural members. The process is not suited for non-ferrous metals. Out-of-position welding is possible with thinner-diameter wire. A variety of wires bearing the prefix "FC" are available in diameters of 3/64 to 1/8 inch. The basic difference between them is the composition of the flux contained within the external sheath. This includes the type and amount of deoxidizers, scavengers, and metallic elements to improve certain mechanical properties of the deposited weld. If it is deemed necessary to use a shield in addition to the flux core, the usual choice is CO_2, mainly because of its low cost. C-

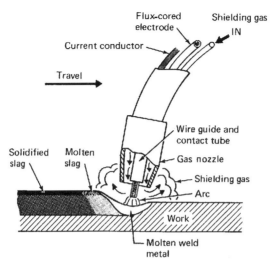

FIG. 16-2 The flux cored arc welding process. (Courtesy of Lincoln Electric Co.)

25 (75 percent Ar-25 percent CO2) and argon-oxygen mixtures, combined with specific wires, can also be used where extra high quality and superior appearance is required.

Welding Technique

Basically the same as GMAW, with similar preselected and welder-determined variables. Welding is done using the forehand method with the gun pointed approximately 20 degrees in the direction of travel. Welding speeds tend to be higher than GMAW and wire stick out distances are longer than those used in GMAW, typically 1 to 1-1/4 inches. The process is best suited for in-position welding. However, out-of-position welding can be accomplished usually with a wire diameter of no more than 0.045 inch and a C-25 gas mixture is used. Some torch manipulation may be required to avoid slag entrapment. The need for this movement is generally reduced with the use of a gas shield. Additionally, some situations may allow multipass pass welding without the need to remove the slag between succssive beads. When the flux core alone is used, slag chipping between passes is required as in conventional welding. FCAW produces more smoke and fumes than any other manual process. Curved front helmets and adequate ventilation systems are recommended. Additional FCAW data can be found in the appendix.

Summation of Welding Settings (GMAW and FCAW)

Initial values for volts, amperes, and shielding gas flow rates can be obtained from data sheets in the appendices. It must be understood that these are only "ballpark" settings based on standard-size test plates welded in various positions. Therefore, it is often necessary to adjust the equipment to suit not only actual job conditions but welder preference and experience as well. The best

procedure is to make test welds on representative sample joints and when optimum welding conditions are obtained, record the settings for future reference.

NOTE: *Welding amperage is a direct result of wire feed adjustment. As wire
speed is increased or decreased, amperage will also increase or decrease.*

Factors to Consider in Wire Selection

The selection of a wire type and its use in combination with a particular
type of shielding gas is most important to successful MIG welding. The wire
selection is made with reference to the following factors:

- Base plate metallurgy
- Plate thickness and joint design
- Shielding gas employed
- Desired mechanical and chemical properties in the weld deposit
- Base plate surface conditions

MIG Wire Classifications

- Steel wire: AWS designation ERXXS-X, with sequential dash
  numbers indicating particular wire chemistry and shielding gas
  requirements. Example: ER80S-132
- Aluminum wire: Numerical designations as with TIG wire.
  Example: ER4043
- Stainless wire: Numerical designations as with TIG wire.
  Example: E308
- Copper and copper alloys: Designated "ECu" plus chemical
  symbol suffix. Example: ECuAL-A2
- Flux core wires: Designated "FC" XXX. Example: FCE-70T5-A1

A compilation of MIG wire recommendations can be found in the appendix.

General Selection Guide

- Steel wires are selected on the basis of bead shape, puddle vis
  cosity, amount of deoxidizers to limit porosity, and carbon or
  alloy content to improve weld deposit metallurgy.
- Aluminum wires are selected on the basis of the heat treatability
  of the deposit, maximum corrosive resistance, and color match
  of deposit during finishing processes.

- Stainless wires are selected on the basis of corrosive resistance of the deposit, minimum carbon content, and expected high-temperature service.
- Copper wires are selected on the basis of weld deposit conductivity required, allowable porosity, and the strength required.

Shielding Gases

Several types of gases are used as shields in MIG welding. These gases are either used singly or in combinations.

Typical Gas Types and Combinations

- Argon (Ar)
- Helium (He)
- Carbon dioxide (CO2)
- Argon/oxygen (Ar-O2)
- Argon/carbon dioxide (Ar-CO2)
- Argon/helium (Ar-He)
- Helium/argon/carbon dioxide (He-Ar-CO2)

The gases used have basic properties that affect the performance of the MIG process. They include:

- Thermal conductivity
- Chemical reaction to the alloying elements in both the workpiece and the filler wire
- Effect of each gas on the mode of metal transfer across the arc

Thermal Conductivity

As thermal conductivity increases, greater welding voltage is required to maintain a stable arc. For example, the thermal conductivity of helium and carbon dioxide is higher than that of argon. Therefore, more heat is delivered to the weld when using those gases.

Chemical Reaction

The compatibility of each gas with both the filler wire and the base metal will determine levels of arc stability, porosity, and wetting action between wire and workpiece. For example, oxygen-bearing shielding gases will cause excessive porosity when welding nonferrous metals. However, certain amounts of oxygen and CO_2 are useful, even essential, in welding steel.

Mode of Metal Transfer

The shielding gas determines either "short" or "spray" arc conditions. The type of gas will also dictate the depth to which the workpiece is melted (penetration). For example, a desired spray arc condition cannot be obtained with a gas rich in CO_2, regardless of voltage levels. However, in short arc welding, CO_2 will provide rather deep penetration and narrow beads, both of which would be ideal in welding outside corner beads on medium-thickness material. Argon is essential for creating a spray arc condition, and argon-enriched mixtures improve bead appearance and reduce spatter in short arc welding.

Setup Variables

After selecting a suitable gas and wire combination, the proper operating conditions must be set up on the equipment

Primary Operating Adjustments Prior to Starting to Weld

- ♦ **Slope**: steep for short arc, flat for spray arc.
- ♦ **Voltage:** refer to appendix for initial settings
- ♦ **Induction:** set at midpoint on dial
- ♦ **Wire speed:** set knob on torch or dial on wire feeder slightly above the midpoint of the range.

Initiating the Arc

- ♦ Select a piece of scrap material similar in size to the production weldment.
- ♦ Hold the torch in the bare hand with a finger on the trigger or start button.
- ♦ Place the gloved opposite hand behind the nozzle and forward of the trigger and use it as a brace to steady the torch and establish a nozzle-to-work distance of approximately 1/2 in.
- ♦ Press the trigger and begin welding.

Stabilizing the Arc

♦ As the wire strikes the work, it may push back against the gun. Ignore this and hold the gun firmly in position.
♦ Look for the establishment of the proper arc length of approximately 1/8 inch. If arc length is longer, readjust the wire feed slightly higher or slightly lower the voltage. If the arc length appears too short and spatter seems excessive, raise the voltage or slow down the wire feed.
♦ If possible, have someone read out the voltage and amperage from the power supply meters. Make adjustments as needed or from the data sheets for the material being welded.
♦ Run beads several inches long and assess their size, shape, penetration, and overall appearance. Adjust induction as needed.

Once these setup variables have proven accurate and optimum welding conditions achieved, they should be recorded for reference when each job is repeated.

Procedure Variables

The preceding primary adjustments and preselected variables determine elements of bead size, depth of penetration, overall welding speed, and method of metal transfer to the puddle. The manipulative variations exercised by the welder will now govern the finer aspects of the process and result in the welding of sound, well-formed beads. The manipulative variables are:

♦ Torch travel speed
♦ Direction of torch movement, forehand or backhand
♦ Torch angle in relation to plate surface
♦ Nozzle-to-work distance (stickout)
♦ Weave patterns

Torch Travel Speed

Simply stated, the faster the torch is moved along the joint, the narrower the bead. A slow-moving torch will form a wider bead. However, in practical welding, where the joint may chill suddenly due to its being adjacent to a large gusset or other mass, or the location of a joint in respect to the whole weldment, such as being near the edge or close to large holes and cutouts causes rapid overheating of the joint, the welder must speed up or slow down accordingly. This control will be based upon the welder's visual estimation of what is needed to fill the joint properly, avoid undercutting, or excessive build up, and maintain an adequate depth of fusion.

Backhand or Forehand Torch Movement

Although the forehand method (Fig. 16-3) usually results in higher welding speeds, many welders find that the backhand method (Fig. 16-4) provides better arc stability and leaves less spatter on the work.

The forehand method produces less penetration and generally narrower beads, which is desirable on thinner material. Welding backhand produces the opposite effect and so would be better suited to thicker material. It should be noted that it may be difficult to follow a poorly defined joint visually while welding backhanded. Because shielding gas coverage is more critical, when MIG welding aluminum, the forehand method is always used.

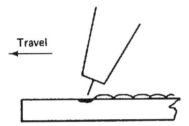

FIG. 16-3 Forehand welding

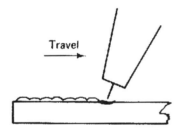

FIG. 16-4 Backhand welding.

Torch Angle

Torch angle is specified in two directions. First, longitudinally along the joint, then transversely in relation to the surface of the plates. Ideally, the torch is held from 5 to 20 degrees off perpendicular longitudinally and 90 degrees transversely on plates being welded in the flat position (Fig. 16-5).

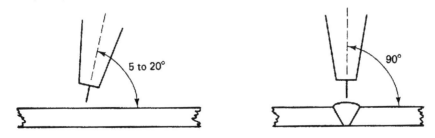

FIG. 16-5 Flat position welding torch angles.

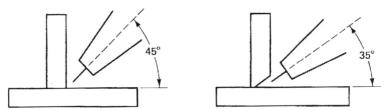

FIG. 16-6 An angle of less than 45 degrees will help minimize undercutting the vertical member of a "tee" joint.

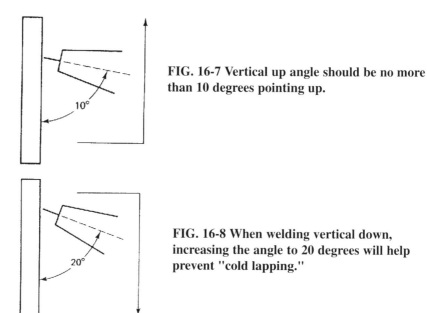

FIG. 16-7 Vertical up angle should be no more than 10 degrees pointing up.

FIG. 16-8 When welding vertical down, increasing the angle to 20 degrees will help prevent "cold lapping."

On tee joints, the torch is held as shown or somewhat less than 45 degrees if the vertical member is beveled (Fig. 16-6). On multiple-pass groove welds, the torch is held 5 to 15 degrees off perpendicular in order to fuse the sides of the groove adequately. In both forehand and backhand welding, the longitudinal angle of the torch should not exceed 25 degrees or some spattering and loss of shielding will result. In welding vertical-up, the longitudinal angle should be no more than 10 degrees (Fig. 16-7), or if welding vertical down, about 20 degrees (Fig. 16-8). Care always must be exercised to avoid cold lapping by making sure that the wire keeps melting into the leading edge of the puddle.

Stickout Distance

The stickout and nozzle-to-work distance are extremely important in determining bead appearance and, to some extent, controlling penetration. As the nozzle-to-work distance increases and the wire stickout becomes longer, there is a decrease in voltage across the arc. Decreasing the nozzle-to-work distance decreases the stickout and causes an increase in arc voltage. Initial voltage settings should be made to allow a 3/8-inch stickout dimension in short-arc welding and a 1/2 to 3/4 inch stickout in spray-arc. The welder should try to maintain these dimensions throughout the weld for the most consistent results, but may vary them slightly to compensate for variations in joint fit-up and geometry. Excessive variation will result in either dirty, porous welds if too long, or clogged nozzles and burned contact tips if held too short. In general, flux-cored welding uses longer stickout distances.

Weave Patterns

Weave patterns in MIG welding are similar to those used in stick welding except for the making of fillet welds when circular motions are recommended. As the deposition rate in MIG welding is higher than in stick welding, care must be exercised to avoid bead pileup and rollover while weaving.

Final Recommendations

The most useful tool in MIG welding is a pair of diagonal cutting pliers. Wire stickout should be cut off flush to the nozzle before restarting and one jaw of the cutters makes an excellent nozzle cleaning tool. The use of an antispatter compound is also recommended to retard spatter buildup inside the nozzle and on the work surfaces.

GMAW Problems and Causes

Defect/Problem	*Cause*
Wire burnback (fusing to contact tip)	1. Initial voltage too high 2. Initial wire feed setting too low 3. Nozzle-to-work distance too long 4. Constriction in wire conduits
Erratic wire feed	1. Improper adjustment of wire drive rolls 2. Improper operation of wire feeder drive and brake system 3. Contact tip too short 4. Kinks in wire conduits
Poor weld starts or wire stubbing	1. Voltage too low 2. Induction and/or slope too high 3. Wire stickout too long
Dirty welds	1. Inadequate shielding gas flow rates 2. Excessive torch angle to work 3. Backhand welding on aluminum 4. Spatter buildup in nozzle 5. Excessive torch weaving
Porosity	1. Dirty base metal 2. Improper wire chemistry 3. Inadequate gas shielding 4. Entrapment of "glass" from previous beads (steel) 5. Short arc welding on aluminum over 1/8 in. thick
Lack of fusion	1. Voltage and/or current too low 2. Welding speed too slow 3. Weld joint too narrow 4. Failure to concentrate arc on leading edge of puddle
Poor bead formation	1. Voltage too high 2. Current too low 3. Induction improperly adjusted 4. Improper torch manipulation
Excessive spatter	1. Use of straight CO_2 (consider Ar-CO_2 or AR-O_2 in steel welding) 2. If welding aluminum, use less or eliminate helium, use straight argon 3. Voltage too low and/or current too high 4. Raise induction and/or slope 5. Lengthen nozzle to work distance
Undercutting	1. Travel speed too fast 2. Voltage too high 3. Welding with excessive current 4. Insufficient pause at edges of weld
Burn-through	1. Current too high 2. Travel too slow 3. Excessive root openings 4. Use of straight CO_2 in steel welding or straight helium in aluminum welding
Cracking	1. Incorrect filler wire selection 2. Beads too small 3. Poor plate quality 4. Rapid chilling of base material

TABLE 16-1 GMAW problems and remedies.

Plasma Welding and Cutting (PAW-PAC)

The Plasma Process

Plasma is referred to as the fourth state of matter after solids, liquids, and gases. The process principle lies in high velocity, ionized gas particles impinging upon the work and generating the heat required for fusion and cutting.

The two forms of plasma technology are the non-transferred arc and the transferred arc (Fig. 17-1). In it's non-transferred form the current flow is between the electrode within the torch and the plasma gas nozzle. This form is used for the spray metalizing processes. The transferred arc method establishes the arc between the electrode, through the work and back to the power supply. This is the form that is used for welding and cutting and where the similarity to direct current, straight polarity GTAW begins.

As a manual welding process, plasma offers particular advantages. The GTAW arc is conical, but the plasma arc is more columnar allowing a more tolerable variation in arc length. This enables the welder to better observe and control of the weld. Faster welding speeds and a narrower heat-affected zone are primary advantages. Most manual applications of plasma arc occur in the low amperage range of 100 amps or less. Because the process can operate at very low-

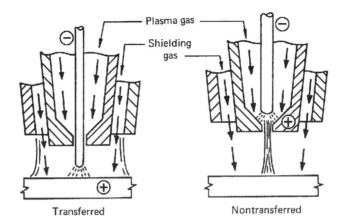

Transferred   Nontransferred

FIG. 17-1 Plasma arc variations. (Courtesy of Lincoln Electric Co.)

amperage, the fusion of foil thickness material is possible. When welding at higher amperage the process is usually automated in some way. The plasma process is applicable to all commercial metals, but finds its greatest use in small, thin weldments of aluminum and stainless steel. Plasma arc may not be the most economical for general welding but for specialized and delicate joining, it has proved its worth.

Plasma Welding Equipment (Fig. 17-2)

Power supplyies for welding are of the constant current type typically used in TIG welding. Although an AC/DC machine can be used, a straight DC machine is preferred. The power supply should have a wide range of fine current adjustments of from 2 to 250 amps for most plasma operations. There are two gas systems, one for plasma arc ionization and one for shielding. Both gases are usually argon, but argon/helium mixtures can be used. A water cooling system is also needed.

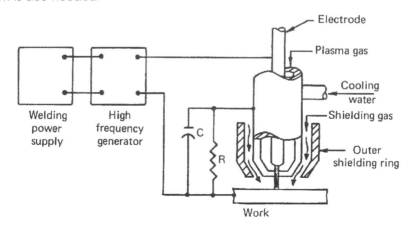

FIG. 17-2 Equipment set-up for PAW. (Courtesy of Lincoln Electric Co.)

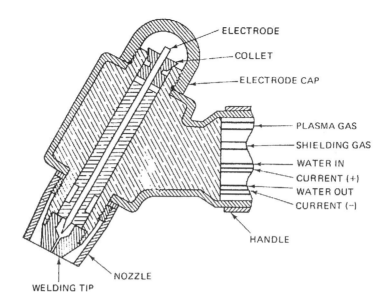

FIG. 17-3 The plasma arc welding torch. (Courtesy of Lincoln Electric Co.)

Plasma welding requires the inclusion of a control console between the torch and the power supply. The console has a separate power source for a pilot arc, systems for initiating the actual plasma arc, water and gas controls and flow meters, and may be connected to a remote contactor. The plasma torch (Fig. 17-3) uses a 2-percent thoriated tungsten which is prepared as if TIG welding steels with DCSP. The torch must be carefully handled and maintained. Centering of the tungsten electrode in the plasma orifice of the nozzle is of particular importance.

Welding Procedure

Procedures and techniques are very similar to those used in TIG welding. Generally welders find that plasma welding is more forgiving and can be performed faster. In addition to more permissible variation in arc length, there is little or no chance of tungsten contamination or spitting into the weld.

Most manual PAW is performed with the "melt-in" method. Very much like a TIG sequence, but much faster, a puddle is formed and filler metal is added as the weld progresses. The melt-in method is used for very thin material, 1/8 inch thick down to less than 0 .010 of an inch thick, and for multipass groove and fillet welds. For thicker material, 1/16 to 1/2 inch, the "key-hole" method is used. Here the plasma jet pierces the joint and the molten metal, supported by its own surface tension, flows around and behind the arc to form the weld bead. Figure 17-4 shows some typical plasma arc joint geometrys and Figure 17-5 details some typical problems and their remedies.

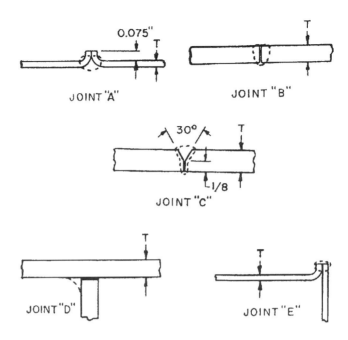

FIG. 17-4 Typical plasma weld joints.

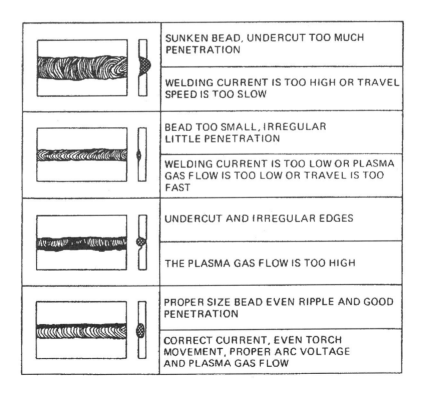

	SUNKEN BEAD, UNDERCUT TOO MUCH PENETRATION
	WELDING CURRENT IS TOO HIGH OR TRAVEL SPEED IS TOO SLOW
	BEAD TOO SMALL, IRREGULAR LITTLE PENETRATION
	WELDING CURRENT IS TOO LOW OR PLASMA GAS FLOW IS TOO LOW OR TRAVEL IS TOO FAST
	UNDERCUT AND IRREGULAR EDGES
	THE PLASMA GAS FLOW IS TOO HIGH
	PROPER SIZE BEAD EVEN RIPPLE AND GOOD PENETRATION
	CORRECT CURRENT, EVEN TORCH MOVEMENT, PROPER ARC VOLTAGE AND PLASMA GAS FLOW

FIG. 17-5 Some plasma weld problems.

Plasma Arc Cutting

In a great many shops the plasma arc process finds its greatest use in cutting and it directly replaces oxy-acetylene as the preferred method. This popularity is due primarily to the ability of the plasma arc to cut stainless and aluminum as well as the carbon steels. Reactive metals such as aluminum and the stainless alloys cannot be cut with oxy-acetylene. The plasma-arc process can be used to cut material as thick as 4 inches, but it is in the manual cutting of sheetmetal thicknesses and lighter plate thicknesses that plasma is most useful. Due to the speed and efficiency of the cut there is negligible distortion, no heat effects on the work, and the quality of the cut often allows assembly and welding with no other preparation. "Stack cutting" is more cleanly done with plasma arc and the process finds wide use in underwater cutting at 100 amps or less. Higher amperages for thicker materials are used with mechanized equipment.

FIG. 17-6 A plasma cutting power supply.

Plasma Arc Cutting Equipment

In general, power supplies with high open-circuit voltages are required, as are the usual gas and water cooling controls, along with contactor control to initiate and stop the cutting process. Lighter duty units may be air cooled. Initially, PAC used nitrogen or argon as the plasma gas and carbon dioxide as the shielding gas and these are still used for quality cuts on aluminum and stainless, 1/2 inch and thicker. Currently, thinner sections of mild steel and aluminum are cut using extra-dry compressed air.

Welding equipment manufacturers today have responded to the increasing popularity of plasma cutting by creating "PAC" equipment packages that include integrated power supplies designed expressly and exclusively for cutting, along with connecting hoses and torches (Fig. 17-6).

Cutting Procedure

Manual cutting is accomplished in very much the same way as using an oxy-acetylene cutting torch (Fig. 17-7). A constant nozzle height above the work is required for smooth cuts. Because the plasma-arc torch is simpler in design, it can be held and guided with more dexterity and accuracy than with oxy-acetylene. From an operative standpoint, skill and care in maintaining the torch and its components are the primary elements in cutting efficiency and quality. The noise generated by the process may call for the hearing protection devices. Generally, welding helmets with either a #9 or #10 lens provide adequate eye protection from the plasma arc glare.

FIG. 17-7 PAC is ideal for cutting sheetmetal thickness' as well as plate material.

SECTION **3**

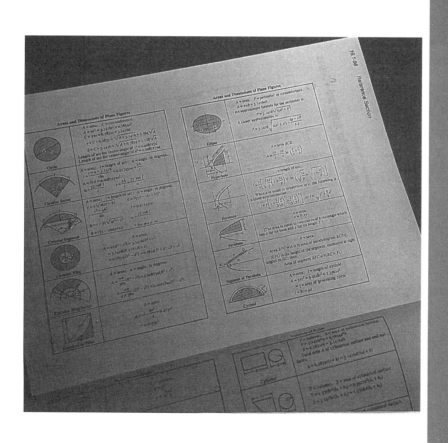

List of Appendices

ASTM Number	NOMINAL COMPOSITION %				Melting Temp		Freezing Temp		Uses
	Tin	Lead	Antimony	Silver	C	F	C	F	
1.55	1	97.5	0.40 max.	1.5	309	588	309	588	For use on copper, brass & similar metals.
2.55	0	97.5	0.40 max.	2.5	304	579	304	579	For use on copper, brass & similar metals. Not in humid environments
5A	5	95	0.12 max.	—	312	594	270	518	For coating & joining metals.
10B	10	90	0.50 max.	—	299	570	268	514	For coating & joining metals
15B	15	85	0.50 max.	—	288	550	227	440	For coating & joining metals
20B	20	80	0.50 max.	—	277	531	183	361	For filling dents or seams in auto bodies
25A	25	75	0.25 max.	—	266	511	183	361	For machine & torch soldering
30A	30	70	0.25 max.	—	255	491	183	361	For machine & torch soldering
35A	35	65	0.25 max.	—	247	477	183	361	General purpose & wiper solder
40A	40	60	0.12 max.	—	238	460	183	361	Wiping solder for auto radiator cores
45A	45	55	0.12 max.	—	227	441	183	361	For auto radiator cores & roofing seams
50A	50	50	0.12 max.	—	216	421	183	361	For general purpose most popular of all
60A	60	40	0.12 max.	—	190	374	183	361	Fine solder where the temp. requirements are critical
63A	63	37	0.12 max.	—	183	361	183	361	As lowest melting (eutectic) solder
70A	70	30	0.12 max.	—	192	378	183	361	For coating metals
20C	20	79	1.0	—	270	517	184	363	For machine soldering & coating of metals, tipping, & like uses
25C	25	73.7	1.3	—	263	504	184	364	For torch & machine soldering
30C	30	68.4	1.6	—	250	482	185	364	For torch soldering or machine soldering
35C	35	63.2	1.8	—	243	470	185	365	For wiping & all uses
40C	40	58	2.0	—	231	448	185	365	Same as (50-50) tin-lead
95TA	95	—	5.0	—	240	464	234	452	For joints on copper, electrical plumbing & heating

Note: The "C" Grades should not be used on galvanized steel.

A-1 Soft solder data

Fraction, Decimal, and Millimeter Conversion Chart

Decimals to Millimeters

Decimal	mm	Decimal	mm
0.001	0.0254	0.500	12.7000
0.002	0.0508	0.510	12.9540
0.003	0.0762	0.520	13.2080
0.004	0.1016	0.530	13.4620
0.005	0.1270	0.540	13.7160
0.006	0.1524	0.550	13.9700
0.007	0.1778	0.560	14.2240
0.008	0.2032	0.570	14.4780
0.009	0.2286	0.580	14.7320
0.010	0.2540	0.590	14.9860
0.020	0.5080	0.600	15.2400
0.030	0.7620	0.610	15.4940
0.040	1.0160	0.620	15.7480
0.050	1.2700	0.630	16.0020
0.060	1.5240	0.640	16.2560
0.070	1.7780	0.650	16.5100
0.080	2.0320	0.660	16.7640
0.090	2.2860	0.670	17.0180
0.100	2.5400	0.680	17.2720
0.110	2.7940	0.690	17.5260
0.120	3.0480	0.700	17.7800
0.130	3.3020	0.710	18.0340
0.140	3.5560	0.720	18.2880
0.150	3.8100	0.730	18.5420
0.160	4.0640	0.740	18.7960
0.170	4.3180	0.750	19.0500
0.180	4.5720	0.760	19.3040
0.190	4.8260	0.770	19.5580
0.200	5.0800	0.780	19.8120
0.210	5.3340	0.790	20.0660
0.220	5.5880	0.800	20.3200
0.230	5.8420	0.810	20.5740
0.240	6.0960	0.820	20.8280
0.250	6.3500	0.830	21.0820
0.260	6.6040	0.840	21.3360
0.270	6.8580	0.850	21.5900
0.280	7.1120	0.860	21.8440
0.290	7.3660	0.870	22.0980
0.300	7.6200	0.880	22.3520
0.310	7.8740	0.890	22.6060
0.320	8.1280	0.900	22.8600
0.330	8.3820	0.910	23.1140
0.340	8.6360	0.920	23.3680
0.350	8.8900	0.930	23.6220
0.360	9.1440	0.940	23.8760
0.370	9.3980	0.950	24.1300
0.380	9.6520	0.960	24.3840
0.390	9.9060	0.970	24.6380
0.400	10.1600	0.980	24.8920
0.410	10.4140	0.990	25.1460
0.420	10.6680	1.000	25.4000
0.430	10.9220		
0.440	11.1760		
0.450	11.4300		
0.460	11.6840		
0.470	11.9380		
0.480	12.1920		
0.490	12.4460		

Fractions to Decimals to Millimeters

Fraction	Decimal	mm	Fraction	Decimal	mm
1/64	0.0156	0.3969	33/64	0.5156	13.0969
1/32	0.0312	0.7938	17/32	0.5312	13.4938
3/64	0.0469	1.1906	35/64	0.5469	13.8906
1/16	0.0625	1.5875	9/16	0.5625	14.2875
5/64	0.0781	1.9844	37/64	0.5781	14.6844
3/32	0.0938	2.3812	19/32	0.5938	15.0812
7/64	0.1094	2.7781	39/64	0.6094	15.4781
1/8	0.1250	3.1750	5/8	0.6250	15.8750
9/64	0.1406	3.5719	41/64	0.6406	16.2719
5/32	0.1562	3.9688	21/32	0.6562	16.6688
11/64	0.1719	4.3656	43/64	0.6719	17.0656
3/16	0.1875	4.7625	11/16	0.6875	17.4625
13/64	0.2031	5.1594	45/64	0.7031	17.8594
7/32	0.2188	5.5562	23/32	0.7188	18.2562
15/64	0.2344	5.9531	47/64	0.7344	18.6531
1/4	0.2500	6.3500	3/4	0.7500	19.0500
17/64	0.2656	6.7469	49/64	0.7656	19.4469
9/32	0.2812	7.1438	25/32	0.7812	19.8438
19/64	0.2969	7.5406	51/64	0.7969	20.2406
5/16	0.3125	7.9375	13/16	0.8125	20.6375
21/64	0.3281	8.3344	53/64	0.8281	21.0344
11/32	0.3438	8.7312	27/32	0.8438	21.4312
23/64	0.3594	9.1281	55/64	0.8594	21.8281
3/8	0.3750	9.5250	7/8	0.8750	22.2250
25/64	0.3906	9.9219	57/64	0.8906	22.6219
13/32	0.4062	10.3188	29/32	0.9062	23.0188
27/64	0.4219	10.7156	59/64	0.9219	23.4156
7/16	0.4375	11.1125	15/16	0.9375	23.8125
29/64	0.4531	11.5094	61/64	0.9531	24.2094
15/32	0.4688	11.9062	31/32	0.9688	24.6062
31/64	0.4844	12.3031	63/64	0.9844	25.0031
1/2	0.5000	12.7000	1	1.0000	25.4000

A-2 Metric conversions/decimal conversions

AWS-ASTM Electrode Classification	Welding Category	General Characteristics
60,000-psi Minimum Tensile Strength		
E6010	Freeze†	Molten weld metal freezes quickly; suitable for welding in all positions with DC reverse-polarity power; has a low-deposition rate and deeply penetrating arc; can be used to weld all types of joints.
E6011	Freeze†	Similar to E6010, except can be used with AC as well as DC power.
E6012	Follow	Faster travel speed and smaller welds than E6010; AC or DC, straight-polarity power; penetration less than E6010. Primary use is for single-pass welding of thin-gage sheet metal in flat, horizontal, and vertical-down positions.
E6013	Follow	Similar to E6012, except can be used with DC (either polarity) or AC power.
E6027	Fill	Deposition rate high since covering contains about 50% iron powder; primary use is for multipass, deep-groove, and fillet welding in the flat position or horizontal fillets, using DC (either polarity) or AC power.
70,000-psi Minimum Tensile Strength		
E7014	Fill-freeze	Higher deposition rate than E6010; usable with DC (either polarity) or AC power; primary use is for inclined and short, horizontal fillet welds.
E7018	Fill-freeze	Suitable for welding low and medium-carbon steels (0.55% C max) in all positions and types of joints. Weld-metal quality and mechanical properties highest of all mild-steel electrodes; usable with DC reverse polarity or AC power.
E7024	Fill	Higher deposition rate than E7014; suitable for flat-position welding and horizontal fillets.
E7028	Fill	Similar to type E7018; used for welding horizontal fillets and grooved fillet welds in flat position.

* E6020, E7015, and E7016 are not included because of their limited usage. Only electrodes up to 3/16-in. diameter can be used in all welding positions (flat, horizontal, vertical, and overhead).

† When used for welding sheet metal, these electrodes have follow-freeze characteristics.

A-3 Mild steel electrode data (SMAW)(Courtesy of Lincoln Electric Co.)

Base Metal		Service	Covered	Bare Rod or
Wrought	Cast*	Condition	Electrode	Filler Wire
201 202 301 302 304 305 308	CF-8 CF-20	As welded or annealed	E308	ER308
302B		As welded	E309,E310	ER309
303 303Se		As welded or annealed	E312,E309	ER314
304L	CF-3	As welded	E308L,E347	ER308L,ER347
308L		As welded	E308L	ER308L
309	CH20	As welded	E309	ER309
309S			E309,E309Cb	ER309
310	CK-20	As welded	E310	ER310
310S		As welded	E310,E310Cb	ER310
316	CF-8M CF-12M	As welded or annealed	E316,E309Cb†	ER316‡
316L	CF-3M	As welded or stress rel'd.	E316L,E309Cb†	ER316L‡
317	CG-8M	As welded or annealed	E317†	ER317
321 321H		As welded	E347	ER321,ER347
347 347H 348 348H		As welded	E347	ER347
403		Annealed or hardened	E410	ER410
410		As welded	E308,E309,E310	ER308,ER309,ER310
405		Annealed As welded	E430 E308,E309,E310	ER430 ER308,ER309,ER310
420		Annealed or hardened As welded	E420 E308,E309,E310	ER420 ER308,ER309,ER310
430		Annealed As welded	E430 E308,E309,E310	ER430 ER308,ER309,ER310
430Ti		As welded	E430	430Ti§, ER430
431		Annealed or hardened As welded	431§ E308,E309,E310	431§ ER308,ER309,ER310
442		Annealed As welded	442§ E308,E309,E310	442§ ER308,ER309,ER310
446		Annealed As welded	446§ E308,E309,E310	446§ ER308,ER309,ER310

A-4 Stainless steel electrode and bare wires (Courtesy of Lincoln Electric Co.)

Tungsten, Nozzle Size, and Gas Flow Chart (GTAW)

Alloy and Thickness (in.)	Tungsten Diameter	Nozzle Size[a]	Gas Flow (CFH)	
Aluminum				
1/16	1/16	4, 5, 6	15	
1/16-3/16	3/32	6, 7	15-20	CFH is doubled with helium
3/16-1/4	1/8-3/16	7, 8, 10	20	
1/4-1/2	3/16-1/4	8, 10, 12	25-30	
Steels				
1/16	0.040-1/16	4, 5, 6	10	
3/32-1/8	1/16-3/32	4, 5, 6	15	
3/16-1/4	3/32-1/8	6, 7, 8	15-20	
Copper alloys				
1/16	1/16	4, 5, 6	15	
1/8-3/16	3/32-1/8	6, 7, 8	20	Multiple passes recommended
1/4-1/2	1/8	8, 10, 12	20	

[a]Nozzle number refers to 1/16-in. increments. Example: A No. 6 size equals a 6/16- or 3/8-in.-diameter orifice.

A-5 GTAW tungsten, nozzle size and gas flow data

Alloy Type	Filler Rod Number	Remarks
3003	1100	Deposit remains soft and ductile
5052	5356	Deposit most similar to parent material
	5154	Best color match after irridite or annodizing
5086	5556	High corrosion resistance
6061	4043	Deposit heat-treatable for highest strength
	or	
	5356	Best color match

A-6 Aluminum filler rod data

Thickness (in.)	Current, DCSP (amp)	Suggested Rod Size (in.)	Average Welding Speed (ipm)	Argon Flow (cfh)
0.035	100	1/16	12—15	8—10
0.049	100 - 125	1/16	12—18	8—10
0.060	100—140	1/16	12—18	8—10
0.089	140—170	3/32	12—18	8—10
0.125	150—200	1/8	10—12	8—10

A-7 Carbon steel welding data (SMAW) (Courtesy of Lincoln Electric Co.)

Typical Procedures for Semiautomatic Gas-Shielded Flux-Cored Arc Welding
With AWS E70T-1 Electrode and CO_2 Gas

Plate Thickness (in.)	Root Opening (in.)	Passes	Electrode Size (in.)	Current DC+ (amp)	Volts	Average Arc Speed (in./min)	Total Time (hr/ft) of weld	Joint Design
3/8	0	2	.045	180	22	8	0.0500	*
1/2	0	3	.045	180	22	8	0.0750	
1/2	3/32	4	0.045	180	22	8	0.100	*
1	3/32	9	0.045	180	22	5	0.356	
1	1/16	6	0.045	180	22	8	0.149	*
2	1/16	16	0.045	180	2	5	0.627	
1/8	0	1	0.045	180	21	36	0.00555	*
1/2	0	2	0.045	180	21	5	0.0834	
3/16	0	1	5/64	350	28	36	0.00555	
1/2	0	3	7/64	450	20	18	0.0333	

* Vertical

A-8 FCAW welding data (Courtesy of Lincoln Electric Co.)

DATA SHEET 12. GMAW Setup Conditions for Steel: Short and Spray Arc

| Material Thickness (in.) | Wire Diameter (in.) | Short Arc | | Spray Arc | | Gas Flow Rates (CFH) | |
		Volts	Amps	Volts	Amps	C-25[a] (short)	Ar-1%O_2 (CFH)(spray)
1/16	0.030	15-17	100-115			30-35	
1/8	0.035	17-20	130-150	25-27	180-200	35-40	40-45
3/16	0.035	18-21	140-200	25-28	160-220	35-40	40-45
1/4	0.045	20-21	190-220	26-30	190-280	40-45	45-50
3/8	0.045	20-22	200-240			40-45	
3/8	1/16			27-30	290-340		45-50
1/2	0.045	20-22	220-250				
1/2	1/16			27-30	310-360		45-50
3/4	1/16			28-31	330-370		50-55

[a]When straight CO_2 is used, arc voltages will be slightly higher.

A-9 GMAW steel welding data (short/spray)

Material Thickness (in.)	Wire Diameter (in.)	Volts	Amps	Gas Flow (Argon) (CFH)
1/8	0.030	19-21	130-140	30-35
1/8	0.035	19-21	140-150	35
1/4	3/64 (046)	19-21	160-210	40-45
3/8	1/16	21-21	220-270	40-45
1/2	1/16	21-26	220-280	40-50
5/8	1/16-3/32	22-28	230-320	45-50
3/4	3/32	22-30	240-330	45-50
1	3/32-1/8	22-32	240-360	50-60

A-10 Aluminum welding data (GMAW)

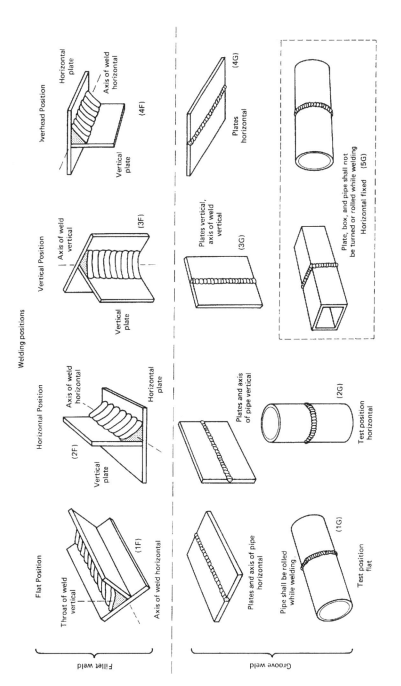

A-11 Welding test positions (Courtesy of Hobart Institute of Welding Technology)

Types of joints

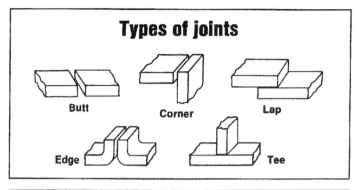

Butt
Corner
Lap
Edge
Tee

Types of welds

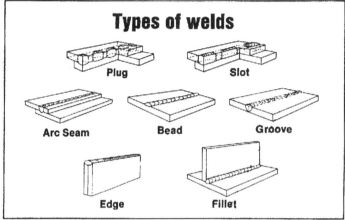

Plug
Slot
Arc Seam
Bead
Groove
Edge
Fillet

Variations of grooves

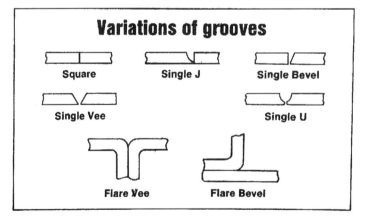

Square
Single J
Single Bevel
Single Vee
Single U
Flare Vee
Flare Bevel

A-12 Weld joint data(Courtesy of Hobart Institute of Welding Technology)

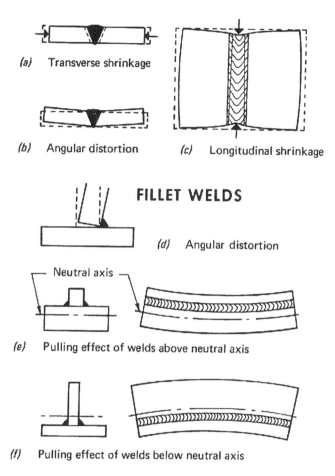

(a) Transverse shrinkage

(b) Angular distortion

(c) Longitudinal shrinkage

FILLET WELDS

(d) Angular distortion

Neutral axis

(e) Pulling effect of welds above neutral axis

(f) Pulling effect of welds below neutral axis

A-13 Weld joint distortion (Courtesy of Lincoln Electric Co.)

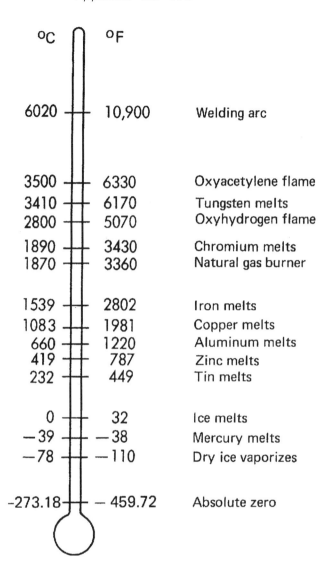

°C   °F

6020 ┼ 10,900   Welding arc

3500 ┼ 6330   Oxyacetylene flame
3410 ┼ 6170   Tungsten melts
2800 ┼ 5070   Oxyhydrogen flame

1890 ┼ 3430   Chromium melts
1870 ┼ 3360   Natural gas burner

1539 ┼ 2802   Iron melts
1083 ┼ 1981   Copper melts
660 ┼ 1220   Aluminum melts
419 ┼ 787   Zinc melts
232 ┼ 449   Tin melts

0 ┼ 32   Ice melts
−39 ┼ −38   Mercury melts
−78 ┼ −110   Dry ice vaporizes

−273.18 ┼ − 459.72   Absolute zero

A-14 Melting temperatures chart

Department of the Air Force

Department of Navy

American Association of State Highway & Transportation Officials

American Bureau of Shipping

American Institute of Steel Construction

American Iron & Steel Construction

American Iron & Steel Institute

American Petroleum Institute

American Society of Mechanical Engineers

American Welding Society

U.S. Department of Transportation

National Certified Pipe Welding Bureau

A-15 Major agencies issuing welding codes and specifications selected

Shielding Gas	Chemical Behavior	Uses, Remarks
Argon	Inert	For welding most metals, except steel
Helium	Inert	Al and Cu alloys, for greater heat and to minimize porosity
A and He (20-80 to 50-50%)	Inert	Al and Cu alloys, for greater heat input and to minimize porosity. Quieter, more stable arc than with He alone.
A and Cl (trace Cl)	Essentially Inert	Al alloys, to minimize porosity
N_2	Reducing	On Cu, permits very powerful arc; used mostly in Europe.
A + 25-30% N_2	Reducing	On Cu, powerful but smoother operating, more readily controlled arc than N_2 alone; used mostly in Europe.
A + 1-2% O_2	Oxidizing	Stainless and alloy steels, also for some deoxidized copper alloys
A + 3-5% O_2	Oxidizing	Plain carbon, alloy, and stainless steels; requires deoxidized electrode
A + 20-30% CO_2	Oxidizing	Various steels; used principally with short-circuiting arc
A + 5% O_2 + 15% CO_2	Oxidizing	Various steels; requires deoxidized wire; used chiefly in Europe
CO_2	Oxidizing	Plain-carbon and low-alloy steels; deoxidized electrode is essential

**A-16 Chemical effects of common shielding gases
(Courtesy of Lincoln Electric Co.)**

Problem	Solution*				
	Current	Voltage	Speed	Stickout	Drag Angle
Porosity	5↑	1↓	4↓	2↑	3↑
Spatter	4↓↑	1↑	5↓	3↓	2↓
Convexity	4↓	1↑	5↓	2↓	3↑
Back Arc Blow	4↓	3↓	5↓	2↑	1↑
Insufficient Penetration	2↑	3↓	4↑	1↓	5↑
Not Enough Follow	4↑	1↓	5↓	2↑	3↑
Stubbing	4↓	1↑		3↓	2↓

* Arrows indicate the need to increase or decrease the setting to correct the problem. Numbers indicate order of importance.
† With E70T-G electrodes, increasing the current reduces droplet size and decreases spatter.

A-17 FCAW trouble shooting chart (Courtesy of Lincoln Electric Co.)

Alloy Type	Recommended Rod	Special Considerations
201, 202, 301, 302	304	Best deposit for low-temperature service
304	308	Highest corrosion resistance in welded area Most commonly fabricated sheet alloy
308, 309, 316, 317, 321, 347	316, 317, or 321	Deposits most similar in mechanical properties to parent material
430, 446	310 or 430	310 leaves more ductile deposit
410, 501, 502 ᵃ	430 or 310	310 most ductile deposit

ᵃAir-hardening types: must be preheated, then annealed after welding.

A-18 Stainless steel filler rod application (GTAW)

Electrode Size	Recommended Current (amp)	
	E3XX-15 Electrodes	E3XX-16 Electrodes
3/32	30—70	30—65
1/8	45—95	55—95
5/32	75—130	80—135
3/16	95—165	120—185
1/4	150—225	200—275
	Optimum current for flat position is about 10% below maximum; optimum for vertical-up welding, about 20% below maximum; optimum for vertical-down welding, about maximum.	Optimum current for flat position is about 10% below maximum; AC range is about 10% higher.

A-19 Current ranges for stainless steel electrodes (SMAW)

GLOSSARY

Glossary of Common Welding Terms

Acetylene: Gas made up of carbon and hydrogen. When mixed with pure oxygen, it will produce one of the hottest gas flame temperatures available.

Alternating current (AC): Electricity that continually reverses its direction of electron flow. In 60-cycle current, this reversal takes place 120 times per second.

Alloy: A mixture of two or more elements.

Annealing: A softening process in which metal is heated to its critical temperature and slowly cooled.

Arc blow: A magnetic disturbance that causes the arc to waver and wander from its intended course during the welding operation.

Arc plasma: A column of ionized gas that conducts electric current to the work.

Arc voltage: Voltage measured across the welding arc.

Arc welding: Fusion welding of metal using an electric arc as the heat source.

Austenitic: A type of stainless steel that is not magnetic and whose basic grain struture is not permanently altered by the heat of welding.

Axis of weld: The line of direction along which the weld is being made.

Backfire: A loud popping sound caused by overheating the oxy-acetylene torch or letting the tip come in contact with the work.

Backhand welding: a welding technique wherein the direction of weld travel is opposite the direction in which the torch is pointing.

Backing strip: A strip of metal positioned on the back side of the weld joint to control penetration.

Back-step welding: Welding small sections of the weld joint in the opposite direction of the overall bead progression.

Base metal: The metal that is to be brazed, welded, or cut.

Bead: The crown or face of the completed weld.

Brittleness: The tendency to break suddenly without bending or stretching.

Braze welding: A method of joining using a bronze alloy filler material and in which some surface alloying takes place

Brazing: Similar to braze welding except that the brazing alloy flows between close-fitting parts by capillary action.

Burnback: The fusing of the MIG welding wire to the contact tube.

Burning: See Flame cutting.

Butt joint: A joint made between two pieces of metal aligned on the same plane.

Cap: The final pass on a weld joint

Capillary action: The ability of soldering or brazing alloys to flow easily into close-fitting spaces.

Carburizing flame: An oxy-acetylene flame that uses more acetylene than oxygen.

Coated electrode: An arc welding electrode that is coated with a flux to aid in the welding operation.

Cold lap: When newly deposited weld metal laps over but does not fuse with weld metal already in place.

Conductivity (thermal): The rate at which heat will travel through a metal.

Contact Tube: A MIG welding component part from which the bare wire electrode picks up current. The tube is sized according to the wire diameter.

Continuous weld: An uninterrupted weld bead that goes from one end of a joint to another or completely around any joint seam.

Crater: A depression at the end of a weld bead.

Crown: The convex surface of the completed weld bead.

Direct current (DC): Electric current that flows in one direction only.

Downhill welding: A vertical weld that progresses from top to bottom.

Duty cycle: The percentage of time out of a ten minute period that a power supply can operate at full output without overheating.

Ductility: The ability to bend, twist, and stretch without cracking.

Electrode: A metal rod or wire that carries electricity to the point where the arc is formed. Most electrodes are of the consumable type, which means that they melt and become part of the weld metal.

Electrode extension: In MIG or flux cored welding, the length of the unmelted electrode extending beyond the contact tube. Also called "stickout."

Expansion (thermal): The amount of dimensional increase in a material as heat is applied.

Face of weld: The exposed top surface of the weld bead.

Ferrous metal: A metal containing iron as its principle ingredient.

Filler rod: A metal rod or wire that is added during the welding operation to produce the desired bead reinforcement.

Fillet weld: An inside corner weld joining two pieces whose surfaces are approximately at 90 degrees to each other.

Fillet weld size: Determined by measurement of the legs of the triangle formed by the fillet.

Flame cutting: Cutting that utilizes the flame of an oxy-acetylene torch and a high pressure stream of oxygen.

Flashback: The regression of the welding flame back into the gas welding torch or cutting device.

Flux: A chemical used to aid adhesion or fusion of metals during a welding or brazing operation.

Forehand welding: A welding technique wherein the torch tip is pointed in the direction of travel.

Fusion: The melting and mixing together of metal.

Globular transfer: Describes the movement of larger drops of weld metal across an arc.

Groove angle: The total or "included" angle between two workpieces.

Groove weld: The welding of two pieces of metal whose joining edges are beveled in order to increase weld penetration.

Hardness: The ability to resist penetration.

Heat-affected zone HAZ: That portion of the base metal which is not melted but is directly affected or altered by the heat of the welding operation.

Heat treatment: The application of heat in specified amounts and for specified time increments in order to affect mechanical properties.

Hot pass: The second pass in a groove weld, usually run at a higher amperage to ensure proper fusion to the root pass.

Inert gas: A gas that will not chemically combine with any weld metal material.

Intermittent weld: The joining of pieces using weld increments rather than a continuous bead.

Joint: The point at which two pieces are to be joined by welding.

Kerf: The width of the cut produced by a cutting operation.

Keyhole: The technique of enlarging a root opening to achieve penetration.

Laser beam: A coherent beam of light used as a heat source for fusion or cutting.

Martensitic: A type of stainless steel whose basic structure will become hard and brittle after welding.

Neutral flame: An oxy-acetylene flame that uses equal amount of oxygen and acetylene.

Nonferrous metal: A metal containing only trace amounts of iron.

Open circuit voltage: The voltage between the output terminals of the power supply with no current flowing.

Orifice: Openings in the ends of welding tips and cutting nozzles through which gas flows.

Oxidizing flame: An oxy-acetylene flame that uses more oxygen than acetylene.

Pass: The weld metal deposited by one progression along the weld joint.

Peening: The mechanical stretching of metals by repeated impact blows.

Penetration: The depth of fusion in the base metal as measured from the base metal surface.

Plug (slot) welding: A method of joining sheets or thinner plates by fusing a nugget of weld material through a hole or slot in one member to the surface of the other.

Porosity: Voids or gas pockets in the weld metal.

Post heating: Any heating performed on the weldment after the welding operation is completed.

Preheating: Any heating operation performed on the base metal prior to welding or cutting.

Pulsed power welding: A procedure which the average current output is exceeded or decreased for a programmed cycle or duration of time.

Puddle: The molten portion of the weld bead.

Purge: To remove the atmosphere by introducing an inert gas.

Reinforcement: Weld metal that extends above the surface of the base metal, either on the face or the root side.

Reverse polarity: Refers to DC welding current, where the electrode is positive and the ground is negative.

Root of weld: The point at which the back side of the weld contacts the base metal surfaces.

Root pass: The first pass in a multipass weld.

Short circuiting transfer: A method of transfer in which weld metal is deposited by a series of short circuits caused by the weld wire striking the work.

Slag inclusion: Nonmetallic material trapped in the weld metal.

Soapstone: A non-dusting chalk used for marking.

Spatter: Unfused globules of weld metal adhering to the surface of the work.

Spray transfer: A method in which fine droplets of weld material are deposited.

Standoff distance: In MIG welding, the distance from the gas nozzle to the work.

Straight polarity: Refers to DC welding current where the electrode is negative and the ground is positive.

Stress relieving: A uniform heating and slow cooling of a weldment for the purpose of relieving internal stresses caused by a welding or cutting operation.

Stringer: A weld bead made with a straight dragging motion with no side-to-side oscillation.

Tack weld: A small weld bead used temporarily to hold parts of a weldment in position for final welding.

Tensile strength: The resistance that a material offers to being pulled apart. It is measured in pounds per square inch.

Tinning: A thin, even coating of the joint area with filler material in a soldering or brazing process.

Toughness: The ability to withstand repeated impact blows without cracking.

Underbead crack: A crack in the heat-affected zone of the weldment and not usually visible on the surface.

Undercut: An unfilled depression or groove melted into the base metal adjacent to the weld bead.

Weave: A weld bead made by manipulating the electrode from side to side.

Welder: One who performs the manipulative acts in welding.

Welding operator: One who operates a welding machine or process equipment

Welding procedure: Specified methods and sequences in completing a welding assignment.

Weldment: An assembly of parts joined by a welding process.

Weld metal: That portion of the base metal and filler rod that is melted and fused together during a welding operation.

Weld tab: Additional joint material on which to begin and end a weld bead.

Weld toe: The boundry between the weld bead and the adjacent work surface.

Yield strength: The point to which a metal can be deformed and not return to its original shape.

I N D E X